T0417900

THE RANDOM MATRIX THEORY OF THE CLASSICAL COMPACT GROUPS

This is the first book to provide a comprehensive overview of foundational results and recent progress in the study of random matrices from the classical compact groups, drawing on the subject's deep connections to geometry, analysis, algebra, physics, and statistics. The book sets a foundation with an introduction to the groups themselves and six different constructions of Haar measure. Classical and recent results are then presented in a digested, accessible form, including the following: results on the joint distributions of the entries; an extensive treatment of eigenvalue distributions, including the Weyl integration formula, moment formulae, and limit theorems and large deviations for the spectral measures; concentration of measure with applications both within random matrix theory and in high dimensional geometry, such as the Johnson–Lindenstrauss lemma and Dvoretzky's theorem; and results on characteristic polynomials with connections to the Riemann zeta function. This book will be a useful reference for researchers in the area, and an accessible introduction for students and researchers in related fields.

ELIZABETH S. MECKES is Professor of Mathematics at Case Western Reserve University. She is a mathematical probabilist specializing in random matrix theory and its applications to other areas of mathematics, physics, and statistics. She received her PhD from Stanford University in 2006 and received the American Institute of Mathematics five-year fellowship. She has also been supported by the Clay Institute of Mathematics, the Simons Foundation, and the US National Science Foundation. She is the author of 24 research papers in mathematics, as well as the textbook *Linear Algebra*, coauthored with Mark Meckes and published by Cambridge University Press in 2018.

CAMBRIDGE TRACTS IN MATHEMATICS

GENERAL EDITORS

A complete list of books in the series can be found at www.cambridge.org/mathematics.
Recent titles include the following:

183. Period Domains over Finite and p-adic Fields. By J.-F. DAT, S. ORLIK, and M. RAPOPORT
184. Algebraic Theories. By J. ADÁMEK, J. ROSICKÝ, and E. M. VITALE
185. Rigidity in Higher Rank Abelian Group Actions I: Introduction and Cocycle Problem. By A. KATOK and V. NIŢICĂ
186. Dimensions, Embeddings, and Attractors. By J. C. ROBINSON
187. Convexity: An Analytic Viewpoint. By B. SIMON
188. Modern Approaches to the Invariant Subspace Problem. By I. CHALENDAR and J. R. PARTINGTON
189. Nonlinear Perron–Frobenius Theory. By B. LEMMENS and R. NUSSBAUM
190. Jordan Structures in Geometry and Analysis. By C.-H. CHU
191. Malliavin Calculus for Lévy Processes and Infinite-Dimensional Brownian Motion. By H. OSSWALD
192. Normal Approximations with Malliavin Calculus. By I. NOURDIN and G. PECCATI
193. Distribution Modulo One and Diophantine Approximation. By Y. BUGEAUD
194. Mathematics of Two-Dimensional Turbulence. By S. KUKSIN and A. SHIRIKYAN
195. A Universal Construction for Groups Acting Freely on Real Trees. By I. CHISWELL and T. MÜLLER
196. The Theory of Hardy's Z-Function. By A. IVIĆ
197. Induced Representations of Locally Compact Groups. By E. KANIUTH and K. F. TAYLOR
198. Topics in Critical Point Theory. By K. PERERA and M. SCHECHTER
199. Combinatorics of Minuscule Representations. By R. M. GREEN
200. Singularities of the Minimal Model Program. By J. KOLLÁR
201. Coherence in Three-Dimensional Category Theory. By N. GURSKI
202. Canonical Ramsey Theory on Polish Spaces. By V. KANOVEI, M. SABOK, and J. ZAPLETAL
203. A Primer on the Dirichlet Space. By O. EL-FALLAH, K. KELLAY, J. MASHREGHI, and T. RANSFORD
204. Group Cohomology and Algebraic Cycles. By B. TOTARO
205. Ridge Functions. By A. PINKUS
206. Probability on Real Lie Algebras. By U. FRANZ and N. PRIVAULT
207. Auxiliary Polynomials in Number Theory. By D. MASSER
208. Representations of Elementary Abelian p-Groups and Vector Bundles. By D. J. BENSON
209. Non-homogeneous Random Walks. By M. MENSHIKOV, S. POPOV and A. WADE
210. Fourier Integrals in Classical Analysis (Second Edition). By C. D. SOGGE
211. Eigenvalues, Multiplicities and Graphs. By C. R. JOHNSON and C. M. SAIAGO
212. Applications of Diophantine Approximation to Integral Points and Transcendence. By P. CORVAJA and U. ZANNIER
213. Variations on a Theme of Borel. By S. WEINBERGER
214. The Mathieu Groups. By A. A. IVANOV
215. Slenderness I: Foundations. By R. DIMITRIC
216. Justification Logic. By S. ARTEMOV and M. FITTING
217. Defocusing Nonlinear Schrödinger Equations. By B. DODSON
218. The Random Matrix Theory of the Classical Compact Groups. By E. S. MECKES

The Random Matrix Theory of the Classical Compact Groups

Cambridge Tracts in Mathematics 218

ELIZABETH S. MECKES
Case Western Reserve University

CAMBRIDGE
UNIVERSITY PRESS

University Printing House, Cambridge CB2 8BS, United Kingdom

One Liberty Plaza, 20th Floor, New York, NY 10006, USA

477 Williamstown Road, Port Melbourne, VIC 3207, Australia

314–321, 3rd Floor, Plot 3, Splendor Forum, Jasola District Centre, New Delhi – 110025, India

79 Anson Road, #06–04/06, Singapore 079906

Cambridge University Press is part of the University of Cambridge.

It furthers the University's mission by disseminating knowledge in the pursuit of education, learning, and research at the highest international levels of excellence.

www.cambridge.org
Information on this title: www.cambridge.org/9781108419529
DOI: 10.1017/9781108303453

First published 2019

Printed in the United Kingdom by TJ International Ltd., Padstow, Cornwall

A catalogue record for this publication is available from the British Library.

Library of Congress Cataloging-in-Publication Data
Names: Meckes, Elizabeth S., author.
Title: The random matrix theory of the classical compact groups /
Elizabeth Meckes (Case Western Reserve University).
Description: Cambridge ; New York, NY : Cambridge University Press, 2019. |
Series: Cambridge tracts in mathematics ; 218 |
Includes bibliographical references and index.
Identifiers: LCCN 2019006045 | ISBN 9781108419529 (hardback : alk. paper)
Subjects: LCSH: Random matrices. | Matrices.
Classification: LCC QA196.5 .M43 2019 | DDC 512.9/434–dc23
LC record available at https://lccn.loc.gov/2019006045

ISBN 978-1-108-41952-9 Hardback

Image by Swallowtail Garden Seeds

To Mark

Contents

Preface

This book grew out of lecture notes from a mini-course I gave at the 2014 Women and Mathematics program at the Institute for Advanced Study. When asked to provide background reading for the participants, I found myself at a bit of a loss; while there are many excellent books that give some treatment of Haar distributed random matrices, there was no single source that gave a broad, and broadly accessible, introduction to the subject in its own right. My goal has been to fill this gap: to give an introduction to the theory of random orthogonal, unitary, and symplectic matrices that approaches the subject from many angles, includes the most important results that anyone looking to learn about the subject should know, and tells a coherent story that allows the beauty of this many-faceted subject to shine through.

The book begins with a very brief introduction to the orthogonal, unitary, and symplectic groups – just enough to get started talking about Haar measure. Section 1.2 includes six different constructions of Haar measure on the classical groups; Chapter 1 also contains some further information on the groups, including some basic aspects of their structure as Lie groups, identification of the Lie algebras, an introduction to representation theory, and discussion of the characters.

Chapter 2 is about the joint distribution of the entries of a Haar-distributed random matrix. The fact that individual entries are approximately Gaussian is classical and goes back to the late nineteenth century. This chapter includes modern results on the joint distribution of the entries in various senses: total variation approximation of principal submatrices by Gaussian matrices, in-probability approximation of (much larger) submatrices by Gaussian matrices, and a treatment of arbitrary projections of Haar measure via Stein's method.

Chapters 3 and 4 deal with the eigenvalues. Chapter 3 is all about exact formulas: the Weyl integration formulas, the structure of the eigenvalue processes as determinantal point processes with explicit kernels, exact formulas due to Diaconis and Shahshahani for the matrix moments, and an interesting decomposition (due to Eric Rains) of the distribution of eigenvalues of powers of random matrices.

Chapter 4 deals with asymptotics for the eigenvalues of large matrices: the sine kernel microscopic scaling limit, limit theorems for the empirical spectral measures and linear eigenvalue statistics, large deviations for the empirical spectral measures, and an interesting self-similarity property of the eigenvalue distribution.

Chapters 5 and 6 are where this project began: concentration of measure on the classical compact groups, with applications in geometry. Chapter 5 introduces the concept of concentration of measure, the connection with log-Sobolev inequalities, and derivations of optimal (at least up to constants) log-Sobolev constants. The final section contains concentration inequalities for the empirical spectral measures of random unitary matrices.

Chapter 6 has some particularly impressive applications of measure concentration on the classical groups to high-dimensional geometry. First, a proof of the celebrated Johnson–Lindenstrauss lemma via concentration of measure on the orthogonal group, with a (very brief) discussion of the role of the lemma in randomized algorithms. The second section is devoted to a proof of Dvoretzky's theorem, via concentration of measure on the unitary group. The final section gives the proof of a "measure-theoretic" Dvoretzky theorem, showing that, subject to some mild constraints, most marginals of high-dimensional probability measures are close to Gaussian.

Finally, chapter 7 gives a taste of the intriguing connection between eigenvalues of random unitary matrices and zeros of the Riemann zeta function. There is a section on Montgomery's theorem and conjecture on pair correlations and one on the results of Keating and Snaith on the characteristic polynomial of a random unitary matrix, which led them to exciting new conjectures on the zeta side. Some numerical evidence (and striking pictures) is presented.

Haar-distributed random matrices appear and play important roles in a wide spectrum of subfields of mathematics, physics, and statistics, and it would never have been possible to mention them all. I have used the end-of-chapter notes in part to give pointers to some interesting topics and connections that I have not included, and doubtless there are many more that I did not mention at all. I have tried to make the book accessible to a reader with an undergraduate background in mathematics generally, with a bit more in probability (e.g., comfort with measure theory would be good). But because the random matrix theory of the classical compact groups touches on so many diverse areas of mathematics, it has been my assumption in writing this book that most readers will not be familiar with all of the background that comes up. I have done my best to give accessible, bottom-line introductions to the areas I thought were most likely to be unfamiliar, but there are no doubt places where an unfamiliar (or, more likely, vaguely familiar, but without enough associations for comfort) phrase will suddenly appear. In these cases, it seems best to take the advice of John von Neumann, who said to a student, "in mathematics you don't understand things. You just get used to them."

One of the greatest pleasures in completing a book is the opportunity to thank the many sources of knowledge, advice, wisdom, and support that made it possible. My thanks first to the Institute for Advanced Study and the organizers of the Women and Mathematics program for inviting me to give the lectures that inspired this book. Thanks also to the National Science Foundation for generous support while I wrote it.

Persi Diaconis introduced me to random matrix theory (and many other things) and taught me to tell a good story.

Amir Dembo encouraged me to embark on this project and gave me valuable advice about how to do it well.

I am grateful to Pierre Albin and Tyler Lawson for their constant willingness to patiently answer all of my questions about geometry and algebra, and, if they didn't already know the answers, to help me wade through unfamiliar literature. Experienced guides make all the difference.

Many thanks to Jon Keating, Arun Ram, and Michel Ledoux for answering my questions about their work and pointing me to better approaches than the ones I knew about. Particular thanks to Nathaël Gozlan for explaining tricky details that eluded me.

My sincere thanks to Andrew Odlyzko for providing the figures based on his computations of zeta zeros.

Thanks to my students, especially Tianyue Liu and Kathryn Stewart, whose questions and comments on earlier drafts certainly enriched the end result.

The excellent and topical photograph for the frontispiece was found (I still don't know how) by Tim Gowers.

As ever, thanks to Sarah Jarosz, this time for *Undercurrent*, which got me most of the way there, and to Yo-Yo Ma for *Six Evolutions*, which carried me to the finish line.

And how to thank my husband and collaborator, Mark Meckes? We have discussed the material in this book for so long and in so many contexts that his viewpoint is inextricably linked with my own. He has lived with the writing of this book, always willing to drop a (probably more important) conversation or task in order to let me hash out a point that suddenly felt terribly urgent. If my writing helps to illuminate the ideas I have tried to describe, it is because I got to talk it out first at the breakfast table.

1

Haar Measure on the Classical Compact Matrix Groups

1.1 The Classical Compact Matrix Groups

The central objects of study in this book are randomly chosen elements of the classical compact matrix groups: the orthogonal group $\mathbb{O}(n)$, the unitary group $\mathbb{U}(n)$, and the symplectic group $\mathbb{S}\mathbb{p}(2n)$. The groups are defined as follows.

Definition

1. An $n \times n$ matrix U over $\mathbb{R}$ is **orthogonal** if

$$UU^T = U^T U = I_n, \tag{1.1}$$

where I_n denotes the $n \times n$ identity matrix, and U^T is the transpose of U. The set of $n \times n$ orthogonal matrices over $\mathbb{R}$ is denoted $\mathbb{O}(n)$.

2. An $n \times n$ matrix U over $\mathbb{C}$ is **unitary** if

$$UU^* = U^* U = I_n, \tag{1.2}$$

where U^* denotes the conjugate transpose of U. The set of $n \times n$ unitary matrices over $\mathbb{C}$ is denoted $\mathbb{U}(n)$.

3. A $2n \times 2n$ matrix U over $\mathbb{C}$ is **symplectic** if $U \in \mathbb{U}(2n)$ and

$$UJU^T = U^T JU = J, \tag{1.3}$$

where

$$J := \begin{bmatrix} 0 & I_n \\ -I_n & 0 \end{bmatrix}. \tag{1.4}$$

The set of $2n \times 2n$ symplectic matrices over $\mathbb{C}$ is denoted $\mathbb{S}\mathbb{p}(2n)$.

Alternatively, the symplectic group can be defined as the set of $n \times n$ matrices U with quaternionic entries, such that $UU^* = I_n$, where U^* is the (quaternionic) conjugate transpose: for

$$\mathbb{H} = \{a + b\mathbf{i} + c\mathbf{j} + d\mathbf{k} : a, b, c, d \in \mathbb{R}\}$$

the skew-field of quaternions, satisfying the relations

$$\mathbf{i}^2 = \mathbf{j}^2 = \mathbf{k}^2 = \mathbf{ijk} = -1,$$

quaternionic conjugation is defined by

$$\overline{a + b\mathbf{i} + c\mathbf{j} + d\mathbf{k}} = a - b\mathbf{i} - c\mathbf{j} - d\mathbf{k}.$$

Quaternions can be represented as 2×2 matrices over $\mathbb{C}$: the map

$$a + b\mathbf{i} + c\mathbf{j} + d\mathbf{k} \longmapsto \begin{bmatrix} a + bi & c + di \\ -c + di & a - bi \end{bmatrix}$$

is an isomorphism of $\mathbb{H}$ onto

$$\left\{ \begin{bmatrix} z & w \\ -\overline{w} & \overline{z} \end{bmatrix} : z, w \in \mathbb{C} \right\}.$$

More generally, if $A, B, C, D \in \mathrm{M}_n(\mathbb{R})$, then the matrix

$$M = A + B\mathbf{i} + C\mathbf{j} + D\mathbf{k} \in \mathrm{M}_n(\mathbb{H})$$

is associated to the matrix

$$M_{\mathbb{C}} = I_2 \otimes A + iQ_2 \otimes B + Q_3 \otimes C + iQ_4 \otimes D,$$

where

$$Q_2 := \begin{bmatrix} 1 & 0 \\ 0 & -1 \end{bmatrix} \qquad Q_3 := \begin{bmatrix} 0 & 1 \\ -1 & 0 \end{bmatrix} \qquad Q_4 := \begin{bmatrix} 0 & 1 \\ 1 & 0 \end{bmatrix}$$

and $\otimes$ denotes the Kronecker product. Any matrix $M \in \mathrm{M}_{2n}(\mathbb{C})$ of this form has the property that

$$MJ = J\overline{M}$$

for $J = Q_3 \otimes I_n$ as above, and the condition $UU^* = I_n$ for $U \in \mathrm{M}_n(\mathbb{H})$ is equivalent to $U_{\mathbb{C}} U_{\mathbb{C}}^* = I_n$ over $\mathbb{C}$.

We will generally consider the symplectic group in its complex version, as a subgroup of the (complex) unitary group, although certain geometric properties of the group can be more cleanly characterized in the quaternionic form.

Note that it is immediate from the definitions that U is orthogonal if and only if U^T is orthogonal, and U is unitary or symplectic if and only if U^* is.

The algebraic definitions given above are nicely compact but may not make the importance of these groups jump right out; the following lemma gives some indication as to why they play such a central role in many areas of mathematics.

Lemma 1.1 *1. Let M be an $n \times n$ matrix over $\mathbb{R}$ or $\mathbb{C}$. Then M is orthogonal or unitary if and only if the columns of M form an orthonormal basis of $\mathbb{R}^n$, resp. $\mathbb{C}^n$.*

2. For U an $n \times n$ matrix over $\mathbb{R}$, $U \in \mathbb{O}(n)$ if and only if U acts as an isometry on $\mathbb{R}^n$; that is,

$$\langle Uv, Uw \rangle = \langle v, w \rangle$$

for all $v, w \in \mathbb{R}^n$.

3. For U an $n \times n$ matrix over $\mathbb{C}$, $U \in \mathbb{U}(n)$ if and only if U acts as an isometry on $\mathbb{C}^n$:

$$\langle Uv, Uw \rangle = \langle v, w \rangle$$

for all $v, w \in \mathbb{C}^n$.

4. Consider $\mathbb{C}^{2n}$ equipped with the skew-symmetric form

$$\omega(v, w) = v_1 w_{n+1} + \cdots + v_n w_{2n} - v_{n+1} w_1 - \cdots - v_{2n} w_n = \sum_{k,\ell} J_{kl} v_k w_\ell,$$

where

$$J = \begin{bmatrix} 0 & I_n \\ -I_n & 0 \end{bmatrix}$$

as above. For a $2n \times 2n$ matrix U over $\mathbb{C}$, $U \in \mathbb{S}\mathbb{p}(2n)$ if and only if U is an isometry of $\mathbb{C}^{2n}$ which preserves ω:

$$\langle Uv, Uw \rangle = \langle v, w \rangle \quad \text{and} \quad \omega(Uv, Uw) = \omega(v, w)$$

for all $v, w \in \mathbb{C}^{2n}$.

5. If $U \in \mathbb{O}(n)$ or $U \in \mathbb{U}(n)$, then $|\det(U)| = 1$. If $U \in \mathbb{S}\mathbb{p}(2n)$, then $\det(U) = 1$.

Proof Note that the $(i, j)^{th}$ entry of $U^T U$ (if U has real entries) or $U^* U$ (if U has complex or quaternionic entries) is exactly the inner product of the ith and jth columns of U. So $U^T U = I_n$ or $U^* U = I_n$ is exactly the same thing as saying the columns of U form an orthonormal basis of $\mathbb{R}^n$ or $\mathbb{C}^n$.

For $U \in \mathrm{M}_n(\mathbb{R})$, $\langle Uv, Uw \rangle = \langle U^T Uv, w \rangle$, and so $\langle Uv, Uw \rangle = \langle v, w \rangle$ for all v and w if and only if $U^T U = I$. The proofs of parts 3 and 4 are similar. For part 5, on any of the groups,

$$|\det(U)|^2 = \det(U)\overline{\det(U)} = \det(U)\det(U^*) = \det(UU^*) = \det(I_n) = 1.$$

The easiest way to see that if $U \in \mathbb{S}\mathbb{p}(2n)$, then in fact $\det(U) = 1$ is to use the Pfaffian: for a skew-symmetric matrix A, the Pfaffian $\mathrm{pf}(A)$ is defined by a sum-over-permutations formula along the lines of the determinant, and has the property that for $2n \times 2n$ matrices A and B,

$$\mathrm{pf}(BAB^T) = \det(B)\,\mathrm{pf}(A).$$

Applying this to the defining relation of $\mathbb{S}\mathbb{p}\,(2n)$,

$$\mathrm{pf}(J) = \mathrm{pf}(UJU^T) = \det(U)\,\mathrm{pf}(J),$$

and so (using the easily verified fact that $\mathrm{pf}(J) \neq 0$), $\det(U) = 1$. □

We sometimes restrict attention to the "special" counterparts of the orthogonal and unitary groups, defined as follows.

Definition The set $\mathbb{SO}\,(n) \subseteq \mathbb{O}\,(n)$ of **special orthogonal matrices** is defined by

$$\mathbb{SO}\,(n) := \{U \in \mathbb{O}\,(n) : \det(U) = 1\}.$$

The set $\mathbb{SO}^-\,(n) \subseteq \mathbb{O}\,(n)$ (the **negative coset**) is defined by

$$\mathbb{SO}^-\,(n) := \{U \in \mathbb{O}\,(n) : \det(U) = -1\}.$$

The set $\mathbb{SU}\,(n) \subseteq \mathbb{U}\,(n)$ of **special unitary matrices** is defined by

$$\mathbb{SU}\,(n) := \{U \in \mathbb{U}\,(n) : \det(U) = 1\}.$$

Since the matrices of the classical compact groups all act as isometries of $\mathbb{C}^n$, all of their eigenvalues lie on the unit circle $\mathbb{S}^1 \subseteq \mathbb{C}$. In the orthogonal and symplectic cases, there are some built-in symmetries:

Exercise 1.2 Show that each matrix in $\mathbb{SO}\,(2n+1)$ has 1 as an eigenvalue, each matrix in $\mathbb{SO}^-\,(2n+1)$ has -1 as an eigenvalue, and each matrix in $\mathbb{SO}^-\,(2n+2)$ has both -1 and 1 as eigenvalues.

The sets $\mathbb{O}\,(n)$, $\mathbb{U}\,(n)$, $\mathbb{S}\mathbb{p}\,(2n)$, $\mathbb{SO}\,(n)$, and $\mathbb{SU}\,(n)$ of matrices defined above are *compact Lie groups*; that is, they are groups (with matrix multiplication as the operation), and they are compact manifolds, such that the multiplication and inverse maps are smooth. Moreover, these groups can naturally be viewed as closed submanifolds of Euclidean space: $\mathbb{O}\,(n)$ and $\mathbb{SO}\,(n)$ are submanifolds of $\mathbb{R}^{n^2}$; $\mathbb{U}\,(n)$ and $\mathbb{SU}\,(n)$ are submanifolds of $\mathbb{C}^{n^2}$; and $\mathbb{S}\mathbb{p}\,(2n)$ is a submanifold of $\mathbb{C}^{(2n)^2}$. Rather than viewing these matrices as n^2-dimensional vectors, it is more natural to view them as elements of the Euclidean spaces $M_n(\mathbb{R})$ (resp. $M_n(\mathbb{C})$) of $n \times n$ matrices over $\mathbb{R}$ (resp. $\mathbb{C}$), where the Euclidean inner products are written as

$$\langle A, B \rangle_{HS} := \mathrm{Tr}(AB^T)$$

for $A, B \in M_n(\mathbb{R})$, and

$$\langle A, B\rangle_{HS} := \mathrm{Tr}(AB^*)$$

for $A, B \in M_n(\mathbb{C})$. These inner products are called the **Hilbert–Schmidt inner products** on matrix space.

The Hilbert–Schmidt inner product induces a norm on matrices; it is sometimes called the Frobenius norm or the Schatten 2-norm, or just the Euclidean norm. This norm is *unitarily invariant*:

$$\|UBV\|_{HS} = \|B\|_{HS}$$

when U and V are unitary (as is easily seen from the definition). This implies in particular that if $U \in \mathbb{O}(n)$ (resp. $\mathbb{U}(n)$), then the map $R_U : M_n(\mathbb{R}) \to M_n(\mathbb{R})$ (resp. $R_U : M_n(\mathbb{C}) \to M_n(\mathbb{C})$) defined by

$$R_U(M) = UM$$

is an isometry on $M_n(\mathbb{R})$ (resp. $M_n(\mathbb{C})$) with respect to the Hilbert–Schmidt inner product.

The Hilbert–Schmidt norm is also *submultiplicative*:

$$\|AB\|_{HS} \leq \|A\|_{HS}\|B\|_{HS}.$$

In fact, this is true of all unitarily invariant norms (subject to the normalization $\|E_{11}\| = 1$), but it is particularly easy to see for the Hilbert–Schmidt norm: let B have columns $b_1, \ldots, b_n$; then $\|B\|_{HS}^2 = \sum_{j=1}^n |b_j|^2$, where $|\cdot|$ is the Euclidean norm on $\mathbb{C}^n$. Now, AB has columns $Ab_1, \ldots, Ab_n$, and so

$$\|AB\|_{HS}^2 = \sum_{j=1}^n |Ab_j|^2 \leq \|A\|_{op}^2 \|B\|_{HS}^2,$$

where $\|A\|_{op} = \sup_{|x|=1} |Ax|$ is the operator norm of A; i.e., the largest singular value of A. Writing the singular value decomposition $A = U\Sigma V$ and using the unitary invariance of the Hilbert–Schmidt norm,

$$\|A\|_{op}^2 = \sigma_1^2 \leq \sum_{j=1}^n \sigma_j^2 = \|\Sigma\|_{HS}^2 = \|A\|_{HS}^2,$$

from which the submultiplicativity follows. Indeed, the sharper estimate

$$\|AB\|_{HS} \leq \|A\|_{op}\|B\|_{HS}$$

is often useful.

The discussion above gives two notions of distance on the classical compact matrix groups: first, the Hilbert–Schmidt inner product can be used to define the distance between two matrices A and B by

$$d_{HS}(A,B) := \|A-B\|_{HS} := \sqrt{\langle A-B, A-B\rangle_{HS}} = \sqrt{\mathrm{Tr}\left[(A-B)(A-B)^*\right]}. \tag{1.5}$$

Alternatively, since, for example, $A, B \in \mathbb{U}(n)$ can be thought of as living in a submanifold of Euclidean space $M_n(\mathbb{C})$, one can consider the *geodesic distance* $d_g(A,B)$ between A and B; that is, the length, as measured by the Hilbert–Schmidt metric, of the shortest path lying entirely in $\mathbb{U}(n)$ between A and B. In the case of $\mathbb{U}(1)$, this is arc-length distance, whereas the Hilbert–Schmidt distance defined in Equation (1.5) is the straight-line distance between two points on the circle. Ultimately, the choice of metric is not terribly important:

Lemma 1.3 *Let $A, B \in \mathbb{U}(n)$. Then*

$$d_{HS}(A,B) \le d_g(A,B) \le \frac{\pi}{2} d_{HS}(A,B).$$

That is, the two notions of distance are equivalent *in a dimension-free way.*

Proof The inequality $d_{HS}(A,B) \le d_g(A,B)$ follows trivially from the fact that the Hilbert–Schmidt distance is the geodesic distance in Euclidean space.

For the other inequality, first note that $d_g(A,B) \le \frac{\pi}{2} d_{HS}(A,B)$ for $A, B \in \mathbb{U}(1)$; that is, that arc-length on the circle is bounded above by $\frac{\pi}{2}$ times Euclidean distance.

Next, observe that both $d_{HS}(\cdot,\cdot)$ and $d_g(\cdot,\cdot)$ are translation-invariant; that is, if $U \in \mathbb{U}(n)$, then

$$d_{HS}(UA, UB) = d_{HS}(A,B) \qquad \text{and} \qquad d_g(UA, UB) = d_g(A,B).$$

In the case of the Hilbert–Schmidt distance, this is immediate from the fact that the Hilbert–Schmidt norm is unitarily invariant. For the geodesic distance, translation invariance follows from the fact that, since any matrix $U \in \mathbb{U}(n)$ acts as an isometry of Euclidean space, every path between A and B lying in $\mathbb{U}(n)$ corresponds to a path between UA and UB of the same length, also lying in $\mathbb{U}(n)$.

Now fix $A, B \in \mathbb{U}(n)$ and let $A^{-1}B = U\Lambda U^*$ be the spectral decomposition of $A^{-1}B$. Then for either distance,

$$d(A,B) = d(I_n, A^{-1}B) = d(I_n, U\Lambda U^*) = d(U^*U, \Lambda) = d(I_n, \Lambda),$$

and so it suffices to assume that $A = I_n$ and B is diagonal.

Write $B = \mathbf{diag}(e^{i\theta_1}, \ldots, e^{i\theta_n})$. Then the length of the path in $\mathbb{U}(n)$ from A to B given by $U(t) := \mathbf{diag}(e^{it\theta_1}, \ldots, e^{it\theta_n})$, for $0 \le t \le 1$ is

$$\begin{aligned}\int_0^1 \|U'(t)\|_{HS}\, dt &= \int_0^1 \left\|\mathbf{diag}(i\theta_1 e^{it\theta_1}, \ldots, i\theta_n e^{it\theta_n})\right\|_{HS} dt \\ &= \int_0^1 \sqrt{\theta_1^2 + \cdots + \theta_n^2}\, dt \\ &\le \frac{\pi}{2}\int_0^1 \sqrt{|1-e^{i\theta_1}|^2 + \cdots + |1-e^{i\theta_n}|^2}\, dt \\ &= \frac{\pi}{2}\left\|I_n - \mathbf{diag}(e^{i\theta_1}, \ldots, e^{i\theta_n})\right\|_{HS},\end{aligned}$$

using the fact that

$$\theta^2 = d_g(1, e^{i\theta})^2 \le \frac{\pi^2}{4} d_{HS}(1, e^{i\theta}),$$

as noted above. □

1.2 Haar Measure

The main goal of this book is to answer the broad general question: What is a random orthogonal, unitary, or symplectic matrix like? To do this, a natural probability measure on each of these groups is needed.

Just as the most natural probability measure (i.e., uniform measure) on the circle is defined by rotation invariance, if G is one of the matrix groups defined in the last section, a "uniform random element" of G should be a random $U \in G$ whose distribution is *translation-invariant*; that is, if $M \in G$ is any fixed matrix, then the equality in distribution

$$MU \stackrel{d}{=} UM \stackrel{d}{=} U$$

should be satisfied. Phrased slightly differently, the distribution of a uniform random element of G should be a translation-invariant probability measure μ on G: for any measurable subset $\mathcal{A} \subseteq G$ and any fixed $M \in G$,

$$\mu(M\mathcal{A}) = \mu(\mathcal{A}M) = \mu(\mathcal{A}),$$

where $M\mathcal{A} := \{MU : U \in \mathcal{A}\}$ and $\mathcal{A}M := \{UM : U \in \mathcal{A}\}$.

It is a theorem due to A. Haar that there is one, and only one, way to do this.

Theorem 1.4 *Let G be any of $\mathbb{O}(n)$, $\mathbb{SO}(n)$, $\mathbb{U}(n)$, $\mathbb{SU}(n)$, or $\mathbb{Sp}(2n)$. Then there is a unique translation-invariant probability measure (called* ***Haar measure****) on G.*

The theorem is true in much more generality (in particular, any compact Lie group has a Haar probability measure). In the most general case the property of left-invariance is not equivalent to that of right-invariance, but in the case of compact Lie groups, left-invariance implies right-invariance and vice versa, so the phrase "translation invariance" will be used in what follows, and will be assumed to include both left- and right-invariance.

Exercise 1.5

1. Prove that a translation-invariant probability measure on $\mathbb{O}(n)$ is invariant under transposition: if U is Haar-distributed, so is U^T.
2. Prove that a translation-invariant probability measure on $\mathbb{U}(n)$ is invariant under transposition and under conjugation: if U is Haar-distributed, so are both U^T and U^*.

Theorem 1.4 is an existence theorem that does not itself provide a description of Haar measure in specific cases. In the case of the circle, i.e., $\mathbb{U}(1)$, it is clear that Haar measure is just (normalized) arc-length. The remainder of this section gives six different constructions of Haar measure on $\mathbb{O}(n)$, with some comments about adapting the constructions to the other groups. For most of the constructions, the resulting measure is only shown to be invariant one one side; the invariance on the other side then follows from the general fact mentioned above that on compact Lie groups, one-sided invariance implies invariance on both sides.

The Riemannian Perspective

It has already been noted that $\mathbb{O}(n) \subseteq \mathrm{M}_n(\mathbb{R})$ and that it is a compact submanifold. It has two connected components: $\mathbb{SO}(n)$ and $\mathbb{SO}^-(n)$, the set of orthogonal matrices U with $\det(U) = -1$. At each point U of $\mathbb{O}(n)$, there is a tangent space $T_U(\mathbb{O}(n))$, consisting of all the tangent vectors to $\mathbb{O}(n)$ based at U.

A map between manifolds induces a map between tangent spaces as follows. Let M_1, M_2 be manifolds and $\varphi : M_1 \to M_2$. If $x \in T_pM_1$, then there is a curve $\gamma : [0, 1] \to M_1$ such that $\gamma(0) = p$ and $\gamma'(0) = x$. Then $\varphi \circ \gamma$ is a curve in M_2 with $\varphi \circ \gamma(0) = \varphi(p)$, and $(\varphi \circ \gamma)'(0)$ is a tangent vector to M_2 at $\varphi(p)$. We take this to be the definition of $\varphi_*(x)$ (it must of course be checked that this gives a well-defined linear map on T_pM_1 for each p).

A Riemannian metric g on a manifold M is a family of inner products, one on the tangent space T_pM to M at each point $p \in M$. The submanifold $\mathbb{O}(n)$ inherits such a metric from $\mathrm{M}_n(\mathbb{R})$, since at each point U in $\mathbb{O}(n)$, $T_U(\mathbb{O}(n))$ is a subspace of $T_U(\mathrm{M}_n(\mathbb{R})) \cong \mathrm{M}_n(\mathbb{R})$. Because multiplication by a fixed orthogonal matrix V is an isometry of $M_n(\mathbb{R})$, the induced map on tangent spaces is also an isometry: if $U \in \mathbb{O}(n)$ with $X_1, X_2 \in T_U(\mathbb{O}(n))$ tangent vectors to $\mathbb{O}(n)$ at U, and $R_V : \mathbb{O}(n) \to \mathbb{O}(n)$ denotes multiplication by a fixed $V \in \mathbb{O}(n)$, then

$$g_{VU}((R_V)_* X_1, (R_V)_* X_2) = g_U(X_1, X_2).$$

On any Riemannian manifold, the Riemannian metric uniquely defines a notion of volume. Since the metric is translation-invariant, the normalized volume form on $\mathbb{O}(n)$ is a translation-invariant probability measure; that is, it is Haar measure.

Since each of the classical compact matrix groups is canonically embedded in Euclidean space, this construction works the same way in all cases.

An Explicit Geometric Construction

Recall that $U \in \mathbb{O}(n)$ if and only if its columns are orthonormal. One way to construct Haar measure on $\mathbb{O}(n)$ is to add entries to an empty matrix column by column (or row by row), as follows. First choose a random vector u_1 uniformly from the sphere $\mathbb{S}^{n-1} \subseteq \mathbb{R}^n$ (that is, according to the probability measure defined by normalized surface area). Take u_1 as the first column of the matrix; by construction, $\|u_1\| = 1$. Now choose u_2 randomly according to surface area measure on

$$\left(u_1^{\perp}\right) \cap \mathbb{S}^{n-1} = \left\{x \in \mathbb{R}^n : \|x\| = 1, \langle x, u_1 \rangle = 0\right\}$$

and let this be the second column of the matrix. Continue in this way; each column is chosen uniformly from the unit sphere of vectors that are orthogonal to each of the preceding columns. The resulting matrix

$$\begin{bmatrix} | & & | \\ u_1 & \dots & u_n \\ | & & | \end{bmatrix}$$

is obviously orthogonal; the proof that its distribution is translation-invariant is as follows.

Observe that if M is a fixed orthogonal matrix, then since

$$M \begin{bmatrix} | & & | \\ u_1 & \dots & u_n \\ | & & | \end{bmatrix} = \begin{bmatrix} | & & | \\ Mu_1 & \dots & Mu_n \\ | & & | \end{bmatrix},$$

the first column of $M\begin{bmatrix} | & & | \\ u_1 & \dots & u_n \\ | & & | \end{bmatrix}$ is constructed by choosing u_1 uniformly from $\mathbb{S}^{n-1}$ and then multiplying by M. But $M \in \mathbb{O}(n)$ means that M acts as a linear isometry of $\mathbb{R}^n$, so it preserves surface area measure on $\mathbb{S}^{n-1}$. That is, the distribution of Mu_1 is exactly uniform on $\mathbb{S}^{n-1}$.

Now, since M is an isometry, $\langle Mu_2, Mu_1\rangle = 0$, and because M is an isometry of $\mathbb{R}^n$, it follows that Mu_2 is *uniformly distributed* on

$$(Mu_1)^\perp \cap \mathbb{S}^{n-1} := \left\{x \in \mathbb{R}^n : |x| = 1, \langle Mu_1, x\rangle = 0\right\}.$$

So the second column of $M\left[u_1 \dots u_n\right]$ is distributed uniformly in the unit sphere of the orthogonal complement of the first column.

Continuing the argument, the distribution of $M\left[u_1 \dots u_n\right]$ is exactly the same as the distribution of $\left[u_1 \dots u_n\right]$; i.e., the construction is left-invariant. It follows by uniqueness that it produces Haar measure on $\mathbb{O}(n)$.

To construct Haar measure on $\mathbb{U}(n)$, one need only draw the columns uniformly from complex spheres in $\mathbb{C}^n$. To get a random matrix in $\mathbb{SO}(n)$, the construction is identical except that there is no choice about the last column; the same is true for $\mathbb{SU}(n)$.

The analogous construction on the representation of elements of $\mathbb{S}\mathbb{p}(2n)$ by $2n \times 2n$ unitary matrices works as follows. For U to be in $\mathbb{S}\mathbb{p}(2n)$, its first column u_1 must lie in the set

$$\{x \in \mathbb{C}^{2n} : \|x\| = 1, \langle x, Jx\rangle = 0\},$$

where J is the matrix defined in (1.4). This condition $\langle x, Jx\rangle = 0$ defines a hyperboloid in $\mathbb{C}^n$ (J is unitarily diagonalizable and has eigenvalues i and $-i$, each with multiplicity n). The set above is thus the intersection of the sphere with this hyperboloid; it is an $(n-2)$-dimensional submanifold of $\mathbb{C}^n$ from which we can choose a point uniformly: this is how we choose u_1. If $n > 1$, one then chooses the second column uniformly from the set

$$\{x \in \mathbb{C}^{2n} : \|x\| = 1, \langle x, u_1\rangle = 0, \langle x, Jx\rangle = 0, \langle x, Ju_1\rangle = 0\};$$

for $n = 1$, one chooses the second column uniformly from

$$\{x \in \mathbb{C}^2 : \|x\| = 1, \langle x, u_1\rangle = 0, \langle x, Jx\rangle = 0\, \langle x, Ju_1\rangle = -1\}.$$

The construction continues: the kth column u_k is chosen uniformly from the intersection of the unit sphere, the hyperboloid $\{x : \langle x, Jx\rangle = 0\}$, and the (affine) subspaces given by the conditions $\langle x, Ju_\ell\rangle = 0$ for $1 \le \ell \le \min\{k-1, n\}$ and $\langle x, Ju_\ell\rangle = -1$ for $n+1 \le \ell < k$ (if $k \ge n+2$). The argument that this construction is invariant under right-translation by an element of $\mathbb{S}\mathbb{p}(2n)$ is similar to the argument above, making use of the fact that for $M \in \mathbb{S}\mathbb{p}(2n)$ given, M is an isometry of $\mathbb{C}^n$ and $MJ = JM$.

A Different Inductive Construction

In this construction, a random element of $\mathbb{O}(n)$ is built up successively from smaller groups. Since it is clear how to choose a random element of $\mathbb{O}(1)$ (flip a coin!), one need only describe how to get a random element of $\mathbb{O}(n)$ from a random element of $\mathbb{O}(n-1)$.

Let U_{n-1} be distributed according to Haar measure on $\mathbb{O}(n-1)$, and let $M \in \mathbb{O}(n)$ be independent of U_{n-1} and have its first column distributed uniformly on $\mathbb{S}^{n-1} \subseteq \mathbb{R}^n$. (The distribution of the remaining columns is irrelevant.) Then define U_n by

$$U_n := M \begin{bmatrix} 1 & 0 \\ 0 & U_{n-1} \end{bmatrix}. \tag{1.6}$$

It is not hard to see that the columns of U_n are distributed as described in the previous approach: It is clear that in the matrix

$$\begin{bmatrix} 1 & 0 \\ 0 & U_{n-1} \end{bmatrix},$$

the second column is uniformly distributed in the orthogonal complement of the first, the third is uniform in the orthogonal complement of the first two, etc. So for a deterministic $M \in \mathbb{O}(n)$, the first column of

$$M \begin{bmatrix} 1 & 0 \\ 0 & U_{n-1} \end{bmatrix}$$

is just m_1 (the first column of M), the second column is uniformly distributed in the orthogonal complement of m_1, etc. Taking M to be random but independent of U_{n-1}, it follows that the distribution of the columns of U_n is exactly as in the previous construction.

The construction works exactly the same way for the unitary and special orthogonal and unitary groups, replacing U_{n-1} by a Haar-distributed element of the next smaller-rank group, and choosing M in the desired group with its first column uniformly in the sphere in $\mathbb{R}^n$ or $\mathbb{C}^n$, as appropriate.

For the symplectic group, the construction is similar, albeit slightly more complicated: let U_{n-1} be Haar-distributed in $\mathbb{S}\mathbb{p}(2(n-1))$, and let $M \in \mathbb{S}\mathbb{p}(2n)$ be independent of U_{n-1} and with its first column m_1 uniform in

$$\{x \in \mathbb{C}^{2n} : \|x\| = 1, \langle x, Jx \rangle = 0\}$$

and its $(n+1)^{st}$ column m_{n+1} uniform in

$$\{x \in \mathbb{C}^{2n} : \|x\| = 1, \langle x, m_1 \rangle = 0, \langle x, Jx \rangle = 0 \, \langle x, Jm_1 \rangle = -1\}.$$

Write U_{n-1} as a 2×2 matrix of $(n-1) \times (n-1)$ blocks:

$$U_{n-1} = \begin{bmatrix} (U_{n-1})_{1,1} & (U_{n-1})_{1,2} \\ (U_{n-1})_{2,1} & (U_{n-1})_{2,1} \end{bmatrix},$$

and define $U_n \in \mathbb{S}\mathrm{p}\,(2n)$ by

$$U_n := M \begin{bmatrix} 1 & 0 & 0 & 0 \\ 0 & (U_{n-1})_{1,1} & 0 & (U_{n-1})_{1,2} \\ 0 & 0 & 1 & 0 \\ 0 & (U_{n-1})_{2,1} & 0 & (U_{n-1})_{2,2} \end{bmatrix}.$$

One can then check that $U_n \in \mathbb{S}\mathrm{p}\,(2n)$ and has its columns distributed as in the previous construction of Haar measure.

The Gauss–Gram–Schmidt Approach

This construction is probably the most commonly used description of Haar measure, and also one that is easy to implement on a computer.

Generate a random matrix X by filling an $n \times n$ matrix, with independent, identically distributed (i.i.d.) standard Gaussian entries $\{x_{i,j}\}$. That is, the joint density (with respect to $\prod_{i,j=1}^{n} dx_{ij}$) of the n^2 entries of X is given by

$$\frac{1}{(2\pi)^{\frac{n^2}{2}}} \prod_{i,j=1}^{n} e^{-\frac{x_{ij}^2}{2}} = \frac{1}{(2\pi)^{\frac{n^2}{2}}} \exp\left\{-\frac{\|X\|_{HS}^2}{2}\right\}$$

(the collection of these random matrices is known as the real Ginibre ensemble). The distribution of X is invariant under multiplication by an orthogonal matrix: by a change of variables, the density of the entries of $Y = MX$ with respect to $\prod dy_{ij}$ is

$$\frac{|\det(M^{-1})|}{(2\pi)^{\frac{n^2}{2}}} \exp\left\{-\frac{\|M^{-1}Y\|^2}{2}\right\} = \frac{1}{(2\pi)^{\frac{n^2}{2}}} \exp\left\{-\frac{\|Y\|^2}{2}\right\},$$

since M^{-1} is an isometry.

That is, the distribution above is translation-invariant, but it is not Haar measure because it does not produce an orthogonal matrix. To make it orthogonal, we use the Gram–Schmidt process. Performing the Gram–Schmidt process commutes with multiplication by a fixed orthogonal matrix M: let x_i denote the columns of X. Then, for example, when the x_1 component is removed from x_2, x_2 is replaced with $x_2 - \langle x_1, x_2 \rangle x_2$. If the result is then multiplied by M, the resulting second column is

$$Mx_2 - \langle x_1, x_2 \rangle Mx_2.$$

If, on the other hand, the multiplication by M is done before applying the Gram–Schmidt algorithm, the result is a matrix with columns $Mx_1, \ldots, Mx_n$. Now removing the component in the direction of column 1 from column 2, the new column 2 is

$$Mx_2 - \langle Mx_1, Mx_2 \rangle Mx_1 = Mx_2 - \langle x_1, x_2 \rangle Mx_2,$$

since M is an isometry.

In other words, if $T : GL_n(\mathbb{R}) \to \mathbb{O}(n)$ is the map given by performing the Gram–Schmidt process ($GL_n(\mathbb{R})$ denotes the group of $n \times n$ invertible matrices over $\mathbb{R}$), then for a fixed orthogonal matrix M,

$$MT(X) = T(MX) \stackrel{d}{=} T(X).$$

That is, the distribution of $T(X)$ is supported on $\mathbb{O}(n)$ and is translation-invariant, meaning that we have once again constructed Haar measure.

Remarks

1. In different terminology, the argument above says that if X is a matrix of i.i.d. standard Gaussian random variables and $X = QR$ is the QR-decomposition obtained via the Gram–Schmidt process, then Q is Haar-distributed on the orthogonal group. But, **WARNING:** The QR-decomposition of a matrix is not uniquely defined, and most computer algebra packages *do not* use the Gram–Schmidt algorithm to produce it. The result of which is that having a computer generate a matrix of i.i.d. standard Gaussian random variables and then returning Q from its internal QR-algorithm will not produce a Haar-distributed matrix; see [81] and [42] for further discussion.
2. The same algorithm as above works over $\mathbb{C}$ or $\mathbb{H}$ to produce a random unitary or symplectic matrix. To produce a random element of $\mathbb{SO}(n)$ or $\mathbb{SU}(n)$, one simply needs a final step: after carrying out the Gram–Schmidt process, multiply the final column by the necessary scalar in order to force the determinant of the matrix to be 1; it is not hard to see that this produces Haar measure on the reduced group.

A Second Gaussian Construction

Again, let X be an $n \times n$ random matrix with i.i.d. standard Gaussian entries. It is easy to see that X has rank n with probability 1, which implies in particular that $X^T X$ is a symmetric rank n matrix, and so

$$X^T X = V^T \mathbf{diag}(d_1, \ldots, d_n) V$$

for some $V \in \mathbb{O}(n)$ and $d_1, \ldots, d_n > 0$; $(X^T X)^{-1/2}$ is then defined by

$$(X^T X)^{-1/2} = V^T \mathbf{diag}(d_1^{-1/2}, \ldots, d_n^{-1/2}) V.$$

Now define the random matrix U by

$$U := X(X^T X)^{-1/2}. \tag{1.7}$$

Then U is orthogonal:

$$U^T U = (X^T X)^{-1/2} X^T X (X^T X)^{-1/2} = I_n,$$

and since for $V \in \mathbb{O}(n)$ fixed, $VX \stackrel{d}{=} X$,

$$U \stackrel{d}{=} VX((VX)^T VX)^{-1/2} = VX(X^T X)^{-1/2} = VU.$$

That is, U is distributed according to Haar measure on $\mathbb{O}(n)$.

Haar Measure on $\mathbb{SO}(n)$ *and* $\mathbb{SO}^-(n)$

The constructions above describe how to choose a uniform random matrix from $\mathbb{O}(n)$, but as we noted above, $\mathbb{O}(n)$ decomposes neatly into two pieces, those matrices with determinant 1 ($\mathbb{SO}(n)$) and those with determinant -1 ($\mathbb{SO}^-(n)$). Theorem 1.4 says that $\mathbb{SO}(n)$ has a unique translation-invariant probability measure; it is clear that this is simply the restriction of Haar measure on $\mathbb{O}(n)$ to $\mathbb{SO}(n)$.

There is also a measure that is sometimes called Haar measure on $\mathbb{SO}^-(n)$, which is the restriction of Haar measure from $\mathbb{O}(n)$ to $\mathbb{SO}^-(n)$. The set $\mathbb{SO}^-(n)$ is of course not a group, but rather a coset of the subgroup $\mathbb{SO}(n)$ in the group $\mathbb{O}(n)$; we continue to use the name "Haar measure on $\mathbb{SO}^-(n)$" because this measure is a probability measure invariant under translation within $\mathbb{SO}^-(n)$ by any matrix from $\mathbb{SO}(n)$. Haar measure on $\mathbb{SO}(n)$ and Haar measure on $\mathbb{SO}^-(n)$ are related as follows: if U is Haar-distributed in $\mathbb{SO}(n)$ and $\widetilde{U}$ is any fixed matrix in $\mathbb{SO}^-(n)$, then $\widetilde{U}U$ is Haar-distributed in $\mathbb{SO}^-(n)$.

Euler Angles

Recall the spherical coordinate system in $\mathbb{R}^n$: each $x \in \mathbb{R}^n$ has spherical coordinates $r, \theta_1, \ldots, \theta_{n-1}$, such that

$$\begin{aligned} x_1 &= r \sin(\theta_{n-1}) \cdots \sin(\theta_2) \sin(\theta_1) \\ x_2 &= r \sin(\theta_{n-1}) \cdots \sin(\theta_2) \cos(\theta_1) \\ &\vdots \\ x_{n-1} &= r \sin(\theta_{n-1}) \cos(\theta_{n-2}) \\ x_n &= r \cos(\theta_{n-1}). \end{aligned}$$

Here,

$$\begin{aligned} &0 \leq r < \infty, \\ &0 \leq \theta_1 < 2\pi, \\ &0 \leq \theta_k \leq \pi, \quad 2 \leq k \leq n. \end{aligned}$$

The spherical coordinates of a point are uniquely determined, except in the cases that some θ_k $(k \geq 2)$ is 0 or π, or $r = 0$.

Spherical coordinates are the basis for the parametrization of $\mathbb{SO}(n)$ by the so-called Euler angles. For $\theta \in [0, 2\pi)$ and $1 \leq k \leq n-1$, let

$$U_k(\theta) := \begin{bmatrix} I_{k-1} & & & \\ & \cos(\theta) & \sin(\theta) & \\ & -\sin(\theta) & \cos(\theta) & \\ & & & I_{n-k-1} \end{bmatrix}.$$

All matrices in $\mathbb{SO}(n)$ can be decomposed as a product of matrices of this type, as follows.

Proposition 1.6 *For any $U \in \mathbb{SO}(n)$, there are angles (called Euler angles) $\{\theta_j^k\}_{\substack{1 \leq k \leq n-1 \\ 1 \leq j \leq k}}$ with $0 \leq \theta_1^k < 2\pi$ and $0 \leq \theta_j^k < \pi$ for $j \neq 1$, so that*

$$U = U^{(n-1)} \cdots U^{(1)},$$

where

$$U^{(k)} = U_1(\theta_1^k) \cdots U_k(\theta_k^k).$$

The Euler angles are unique except if some θ_j^k is 0 or π $(j \geq 2)$.

Proof Observe that the result is vacuous for $n = 1$ and holds trivially for $n = 2$. Suppose then that it holds on $\mathbb{SO}(n-1)$, and let $U \in \mathbb{SO}(n)$. Let

$1, \theta_1^{n-1}, \ldots, \theta_{n-1}^{n-1}$ be the spherical coordinates of Ue_n, where e_j denotes the jth standard basis vector of $\mathbb{R}^n$. Then

$$U_1(\theta_1^{n-1}) \cdots U_{n-1}(\theta_{n-1}^{n-1})e_n = U_1(\theta_1^{n-1}) \cdots U_{n-2}(\theta_{n-2}^{n-1}) \begin{bmatrix} 0 \\ \vdots \\ 0 \\ \sin(\theta_{n-1}^{n-1}) \\ \cos(\theta_{n-1}^{n-1}) \end{bmatrix}$$

$$= \cdots = \begin{bmatrix} \sin(\theta_{n-1}^{n-1}) \cdots \sin(\theta_2^{n-1}) \sin(\theta_1^{n-1}) \\ \sin(\theta_{n-1}^{n-1}) \cdots \sin(\theta_2^{n-1}) \cos(\theta_1^{n-1}) \\ \cdots \\ \sin(\theta_{n-1}^{n-1}) \cos(\theta_{n-2}^{n-1}) \\ \cos(\theta_{n-1}^{n-1}) \end{bmatrix};$$

that is,

$$U_1(\theta_1^{n-1}) \cdots U_{n-1}(\theta_{n-1}^{n-1})e_n = Ue_n,$$

and so

$$\left[U_1(\theta_1^{n-1}) \cdots U_{n-1}(\theta_{n-1}^{n-1})\right]^{-1} U = \begin{bmatrix} \widetilde{U} & 0 \\ 0 & 1 \end{bmatrix} \tag{1.8}$$

for some $\widetilde{U} \in \mathbb{SO}(n-1)$. By the induction hypthesis, $\widetilde{U}$ can be written as

$$\widetilde{U} = U^{(n-2)} \cdots U^{(1)}.$$

By mild abuse of notation we now consider each of the implicit factors $U_\ell(\theta_\ell^k)$, a priori elements of $\mathbb{SO}(n-1)$, to be elements of $\mathbb{SO}(n)$ fixing e_n. The claimed factorization of U follows by multiplying both sides of (1.8) by $U^{(n-1)} := U_1(\theta_1^{n-1}) \cdots U_{n-1}(\theta_{n-1}^{n-1})$. □

Haar measure on $\mathbb{SO}(n)$ can be characterized as a distribution on the Euler angles. Observe first that by a right-to-left version of the column-by-column construction, if $U \in \mathbb{SO}(n)$ is distributed according to Haar measure, then Ue_n is a uniform random point on $\mathbb{S}^{n-1}$. Recall that the uniform probability measure on the sphere $\mathbb{S}^{n-1}$ is given in spherical coordinates by

$$\frac{\Gamma\left(\frac{n}{2}\right)}{2\pi^{n/2}} \sin^{n-2}(\theta_{n-1}) \cdots \sin(\theta_2) d\theta_1 \cdots d\theta_{n-1}.$$

That is, the $\{\theta_k^{n-1}\}_{1 \le k \le n-1}$ subset of the Euler angles of U have density

$$\frac{\Gamma\left(\frac{n}{2}\right)}{2\pi^{n/2}} \sin^{n-2}(\theta_{n-1}^{n-1}) \cdots \sin(\theta_2^{n-1})$$

with respect to $d\theta_1^{n-1} \cdots d\theta_{n-1}^{n-1}$. Now, given Ue_n, the vector Ue_{n-1} is uniformly distributed on the unit sphere in orthogonal complement of Ue_n; equivalently, given $\theta_1^{n-1}, \ldots, \theta_{n-1}^{n-1}$,

$$\left[U_1(\theta_1^{n-1}) \cdots U_{n-1}(\theta_{n-1}^{n-1})\right]^{-1} Ue_{n-1}$$

is distributed uniformly on the $(n-1)$-dimensional unit sphere of $e_n^{\perp}$. Since $\{\theta_k^{n-2}\}_{1 \le k \le n-2}$ are exactly the (angular) spherical coordinates of this vector, it follows that $\{\theta_k^{n-2}\}_{1 \le k \le n-2}$ are independent of $\{\theta_k^{n-1}\}_{1 \le k \le n-1}$, with density

$$\frac{\Gamma\left(\frac{n-1}{2}\right)}{2\pi^{(n-1)/2}} \sin^{n-3}(\theta_{n-2}^{n-2}) \cdots \sin(\theta_2^{n-2}).$$

Continuing in this way, the Euler angles of U are independent, with joint density

$$\prod_{k=1}^{n-1} \left(\frac{\Gamma\left(\frac{k}{2}\right)}{2\pi^{k/2}} \right) \prod_{j=1}^{k} \sin^{j-1}(\theta_j^k). \tag{1.9}$$

From this construction, one can describe a natural analog for $\mathbb{SO}^-(n)$; any $U \in \mathbb{SO}^-(n)$ can be written as

$$U = U^{(n-1)} \cdots U^{(1)} U^{(0)}, \tag{1.10}$$

where

$$U^{(0)} = \begin{bmatrix} -1 & \\ & I_{n-1} \end{bmatrix},$$

and $U^{(n-1)} \cdots U^{(1)}$ is the Euler angle decomposition of $UU^{(0)}$; i.e., the matrix with the same columns as U, except for a sign change in the first column. That is, Haar measure on $\mathbb{SO}^-(n)$ can be described by choosing angles $\{\theta_j^k\}_{\substack{1 \le k \le n-1 \\ 1 \le j \le k}}$ according to the density in (1.9) and then letting U be given by the formula in (1.10).

To choose U according to Haar measure on $\mathbb{O}(n)$, one simply chooses the Euler angles according to (1.9), and then includes the $U^{(0)}$ factor with probability $\frac{1}{2}$, independent of the choice of angles.

1.3 Lie Group Structure and Character Theory

The Classical Groups as Lie Groups

As we have already noted, the classical compact matrix groups are Lie groups; i.e., they are groups and they are differentiable manifolds such that the multiplication operation $(A, B) \mapsto AB$ and the inversion operation $A \mapsto A^{-1}$ are smooth maps. Each of the groups has an associated *Lie algebra*, which is the tangent space to the group at the identity matrix. Lie algebras play an important role in understanding Lie groups, because they are relatively simple objects geometrically (vector spaces!), but they come equipped with an extra algebraic structure called the *Lie bracket*, which encodes much of the geometry of the group itself.

The tool that connects the Lie algebra to the Lie group is the exponential map. While one can define the exponential map on Riemannian manifolds in general, the definition in the setting of the classical compact matrix groups can be made very concrete; this in particular makes the source of the terminology clear.

Definition Let $X \in \mathrm{M}_n(\mathbb{C})$. The matrix exponential e^X is defined by

$$e^X := I_n + X + \frac{1}{2}X^2 + \frac{1}{3!}X^3 + \cdots.$$

We will make frequent use of the following basic facts.

Lemma 1.7 *1. The sum defining the matrix exponential is convergent for any $X \in \mathrm{M}_n(\mathbb{C})$.*
2. If $X \in \mathrm{M}_n(\mathbb{C})$, e^X is invertible with inverse e^{-X}.
3. If X and Y commute, then $e^{X+Y} = e^X e^Y$; this need not be true if X and Y do not commute.

Proof 1. Recall that the Hilbert–Schmidt norm is submultiplicative; from this it follows that

$$\sum_{j=0}^{N} \frac{\|X^j\|_{HS}}{j!} \le \sum_{j=0}^{N} \frac{\|X\|_{HS}^j}{j!} \le \sum_{j=0}^{\infty} \frac{\|X\|_{HS}^j}{j!} < \infty$$

for all N, and so $\sum_{j=1}^{\infty} \frac{1}{j!} X^j$ is convergent.
2. This point follows easily from the next, since X and $-X$ commute.
3. For $N \in \mathbb{N}$, since X and Y commute, we have

$$\begin{aligned}\sum_{j=0}^{N} \frac{1}{j!}(X+Y)^j &= \sum_{j=0}^{N} \frac{1}{j!} \sum_{k=0}^{j} \binom{j}{k} X^k Y^{j-k} \\ &= \sum_{k=0}^{N} \frac{1}{k!} X^k \sum_{j=k}^{N} \frac{1}{(j-k)!} Y^{j-k} = \sum_{k=0}^{N} \frac{1}{k!} X^k \sum_{j=0}^{N-k} \frac{1}{(j)!} Y^j.\end{aligned}$$

Now let $N \to \infty$.

It is easy to cook up examples of noncommuting X and Y for which $e^{X+Y} \neq e^X e^Y$. □

Suppose now that $X = UAU^*$ with $U \in \mathbb{U}(n)$. Then

$$e^X = \sum_{j=0}^{\infty} \frac{1}{j!}(UAU^*)^j = \sum_{j=0}^{\infty} \frac{1}{j!} UA^jU^* = Ue^AU^*.$$

In particular, if X is unitarily diagonalizable (i.e., normal), with $X = U\,\mathbf{diag}(d_1, \ldots, d_n)U^*$, then $e^X = U\,\mathbf{diag}(e^{d_1}, \ldots, e^{d_n})U^*$. More generally, given a function f on $\mathbb{C}$, for unitarily diagonalizable X as above, we *define* $f(X)$ by this route: if $X = U\,\mathbf{diag}(d_1, \ldots, d_n)U^*$, then

$$f(X) := U\,\mathbf{diag}(f(d_1), \ldots, f(d_n))U^*.$$

This prodecure is referred to as the *functional calculus*.

The function $\gamma(t) = e^{tX}$ defines a one-parameter subgroup of $GL_n(\mathbb{C})$ (the group of invertible $n \times n$ matrices over $\mathbb{C}$), since $e^{tX}e^{sX} = e^{(t+s)X}$. More geometrically, $\gamma(t)$ is a curve in $GL_n(\mathbb{C})$ with $\gamma(0) = I_n$ and $\gamma'(0) = X$. In general, the role of the exponential map in Riemannian geometry is that it gives a local diffeomorphism of the tangent space to a manifold at a point to a neighborhood of that point in the manifold. In the present context, it is what maps the Lie algebra of a closed subgroup of $GL_n(\mathbb{C})$ down to a neighborhood of I_n in the subgroup. The following lemma, whose proof we omit, gives the precise statement needed.

Lemma 1.8 *If G is any closed subgroup of $GL_n(\mathbb{C})$, then X is an element of the Lie algebra of G if and only if $e^{tX} \in G$ for all $t \geq 0$. The map $X \mapsto e^X$ gives a diffeomorphism of a neighborhood of 0 in the Lie algebra of G to a neighborhood of I_n in G.*

The lemma allows us to identify the Lie algebras of the classical groups concretely, as follows.

Lemma 1.9

1. *The Lie algebra of $\mathbb{O}(n)$ is*

$$\mathfrak{o}(n) = \left\{ X \in \mathrm{M}_n(\mathbb{R}) : X + X^T = 0 \right\}.$$

2. *The Lie algebra of $\mathbb{SO}(n)$ is*

$$\mathfrak{so}(n) = \mathfrak{o}(n) = \left\{ X \in \mathrm{M}_n(\mathbb{R}) : X + X^T = 0 \right\}.$$

3. *The Lie algebra of $\mathbb{U}(n)$ is*

$$\mathfrak{u}(n) = \left\{ X \in \mathrm{M}_n(\mathbb{C}) : X + X^* = 0 \right\}.$$

4. *The Lie algebra of* $\mathbb{SU}(n)$ *is*

$$\mathfrak{su}(n) = \left\{ X \in \mathrm{M}_n(\mathbb{C}) : X + X^* = 0, \mathrm{Tr}(X) = 0 \right\}.$$

5. *The Lie algebra of* $\mathbb{S}\mathbb{p}(2n) \subseteq \mathbb{U}(n)$ *is*

$$\mathfrak{sp}(2n) = \left\{ X \in \mathrm{M}_{2n}(\mathbb{C}) : X + X^* = 0, XJ + JX^* = 0 \right\}.$$

The quaternionic form of the Lie algebra of $\mathbb{S}\mathbb{p}(2n) \subseteq \mathrm{M}_n(\mathbb{H})$ *is*

$$\mathfrak{su}_{\mathbb{H}}(n) = \left\{ X \in \mathrm{M}_n(\mathbb{H}) : X + X^* = 0 \right\},$$

where X^* *denotes the quaternionic conjugate transpose.*

Proof

1. First, if $\gamma : [0,1] \to \mathbb{O}(n)$ is a curve with $\gamma(0) = I_n$, then $\gamma(t)\gamma(t)^T = I_n$ for each t. Differentiating gives that

$$\gamma'(t)\gamma(t)^T + \gamma(t)\gamma'(t)^T = 0$$

for all t, and so in particular, if $X = \gamma'(0)$ is the tangent vector to γ at I_n, then $X + X^T = 0$.

On the other hand, given X with $X + X^T = 0$,

$$e^{tX}(e^{tX})^T = e^{tX}e^{tX^T} = e^{tX}e^{-tX} = I_n,$$

and so $\gamma(t) = e^{tX}$ is a curve in $\mathbb{O}(n)$ with $\gamma(0) = I_n$ and $\gamma'(0) = X$. That is, X is in the tangent space to $\mathbb{O}(n)$ at I_n.

2. Since $\mathbb{SO}(n)$ is a subgroup of $\mathbb{O}(n)$, the tangent space to $\mathbb{SO}(n)$ at I_n is a subspace of $\mathfrak{o}(n)$. In fact, it is clear geometrically that the Lie algebras of $\mathbb{O}(n)$ and $\mathbb{SO}(n)$ must be the same, since $\mathbb{O}(n)$ is just the union of two disconnected copies of $\mathbb{SO}(n)$ (the second copy being the negative coset $\mathbb{SO}^-(n)$).
3. Exactly analogous to 1.
4. As in the orthogonal case, since $\mathbb{SU}(n)$ is a subgroup of $\mathbb{U}(n)$, the tangent space to $\mathbb{SU}(n)$ at I_n is a subspace of $\mathfrak{u}(n)$; in this case, however, it is not the whole space. The additional condition for a curve $\gamma(t)$ to lie in $\mathbb{SU}(n)$ is that $\det \gamma(t) = 1$ for all t. Using the functional calculus, if $X = U\,\mathbf{diag}(d_1, \dots, d_n)U^*$,

$$\det(e^X) = \det(U\,\mathbf{diag}(e^{d_1}, \dots, e^{d_n})U^*) = e^{d_1 + \cdots + d_n} = e^{\mathrm{Tr}\,X}.$$

In particular, if $\gamma(t) = e^{tX}$, then

$$\frac{d}{dt}(\det(\gamma(t))) = \frac{d}{dt}\left(e^{t\,\mathrm{Tr}\,X}\right) = (\mathrm{Tr}\,X)e^{t\,\mathrm{Tr}\,X},$$

and so X is tangent to $\mathbb{SU}(n)$ at I_n if and only if $X \in \mathfrak{u}(n)$ (i.e., $X + X^* = 0$) and $\operatorname{Tr} X = 0$.

5. For the complex form, since $\mathbb{Sp}(2n)$ is a subgroup of $\mathbb{U}(2n)$, the Lie algebra of $\mathbb{Sp}(2n)$ is a subspace of $\mathfrak{u}(2n)$, hence the first condition in $\mathfrak{sp}(2n)$. For the second, observe that if $\gamma : [0,1] \to \mathbb{Sp}(2n)$ is a curve with $\gamma(0) = I_{2n}$, then differentiating the requirement that $\gamma(t)J\gamma(t)^* = J$ for all t gives that $\gamma'(t)J\gamma(t)^* + \gamma(t)J\gamma'(t)^* = 0$; if $X = \gamma'(0)$, then evaluating at $t = 0$ gives that $XJ + JX^* = 0$.

   On the other hand, if X satisfies $X + X^* = 0$ and $XJ + JX^* = 0$, then

$$e^{tX}J = \sum_{n=0}^{\infty} \frac{t^n}{n!} X^n J = \sum_{n=0}^{\infty} \frac{t^n}{n!} (-X^{n-1}JX^*) = \sum_{n=0}^{\infty} \frac{t^n}{n!} (X^{n-2}J(X^*)^2)$$

$$= \cdots = J \sum_{n=0}^{\infty} \frac{(-tX^*)^n}{n!} = J(e^{-tX})^*,$$

   and so $e^{tX} \in \mathbb{Sp}(2n)$ for all t.

   Verifying the quaternionic form is exactly the same as the unitarycase. □

Observe that even though the matrices in $\mathfrak{u}(n)$, $\mathfrak{su}(n)$, and $\mathfrak{sp}(2n)$ have complex entries, they are *real* vector spaces only; they are not closed under multiplication by complex scalars. They do inherit a real inner product structure from the Euclidean structure of the spaces in which they reside: the inner product on $\mathfrak{u}(n)$, $\mathfrak{su}(n)$, and $\mathfrak{sp}(2n)$ is

$$\langle X, Y \rangle := \operatorname{Re}(\operatorname{Tr}(XY^*)).$$

This is the unique real inner product that defines the same norm as the usual Hilbert–Schmidt inner product on $\mathfrak{u}(n)$, $\mathfrak{su}(n)$, and $\mathfrak{sp}(2n)$.

Representation Theory

We have already encountered two natural actions of the classical compact groups on vector spaces; namely, on $\mathbb{R}^n$ or $\mathbb{C}^n$ by matrix-vector multiplication, and on $M_n(\mathbb{R})$ or $M_n(\mathbb{C})$ by multiplication. The topic of representation theory is to understand *all* the ways a group can act on a vector space. This is a vast and well-developed field, but in the context of the random matrix theory on these groups, our main interest will be not in representation theory in its full glory, but specifically in character theory. Essentially the reason for this is that the irreducible characters form a basis for the space of class functions on a group; i.e., those functions that are constant on conjugacy classes. Conjugacy classes of matrices within the classical groups are exactly the set of matrices with a given set of eigenvalues, and so ultimately, a good basis of the space of class functions gives us a route to studying eigenvalues.

A **representation** of a finite group G is a group homomorphism $\rho : G \to GL(V)$, where V is a finite-dimensional vector space and $GL(V)$ is the group of invertible linear maps on V. A **representation** of a Lie group G is again a map $\rho : G \to GL(V)$ for a finite-dimensional vector space V, which is both a homomorphism of groups and is required to be smooth. That is, a representiation of G is a way of seeing G as acting on V, in a way that respects the structure of G as a (Lie) group. Usually the notation for the map itself is suppressed, and one refers to a representation V of a group G, and writes $g \cdot v$ or just gv rather than $\rho(g)(v)$.

A representation V of G is **irreducible** if V has no proper nontrivial subspaces which are invariant under the action of G.

Example Let S_n be the symmetric group on n letters; S_n acts on $\mathbb{R}^n$ by permuting the coordinates of a vector. This representation of S_n is not irreducible, since the subspace V_1 spanned by $(1, 1, \ldots, 1)$ is invariant under the action of S_n, as is the complementary subspace $V_2 = \{(x_1, \ldots, x_n) : x_1 + \cdots + x_n = 0\}$. Of course, V_1 is an irreducible representation of S_n (it is called the trivial one-dimensional representation). Less obviously, though not too hard to see, V_2 is also an irreducible representation of S_n; it is called the *standard representation*.

Known representations of a group give rise to new ones in various ways. The simplest are by taking direct sums or tensor products: if V and W are representations of G, then $V \oplus W$ and $V \otimes W$ are also representations of G, via the actions

$$g((v, w)) = (gv, gw) \qquad g(v \otimes w) = (gv) \otimes (gw).$$

A representation V also induces a representation on the dual space V^* of scalar-valued linear functions on V: if $v^* \in V^*$, then

$$(gv^*)(v) := v^*(g^{-1}v).$$

A linear map $\varphi : V \to W$ between representations V and W of a group G is called G-linear if for all $v \in V$ and all $g \in G$,

$$g \cdot \varphi(v) = \varphi(g \cdot v).$$

Two representations V and W are **isomorphic** if there is a G-linear isomorphism between the vector spaces V and W.

A fundamental fact is that finite-dimensional representations of finite groups and compact Lie groups can be decomposed into direct sums of irreducible representations. The proof in either case is essentially the same, and rests on the fact that for any representation V of such a group, one can define an inner product on V such that each group element acts as an isometry. Indeed, take any inner product $\langle \cdot, \cdot \rangle$ on V, and define

$$\langle v, w \rangle_G := \begin{cases} \frac{1}{|G|} \sum_{g \in G} \langle gv, gw \rangle, & G \text{ finite;} \\ \int \langle gv, gw \rangle \, dg, & G \text{ a compact Lie group,} \end{cases}$$

where the integration in the second case is with respect to normalized Haar measure on G. Then if $W \subsetneq V$ is a nonzero subspace that is invariant under the action of G, the orthogonal complement $W^\perp$ of W with respect to $\langle \cdot, \cdot \rangle_G$ is also G-invariant, and $V = W \oplus W^\perp$. Continuing in this way defines a decomposition.

Suppose now that $V = \oplus_{i=1}^k V_i^{\oplus a_i} = \oplus_{j=1}^\ell W_j^{\oplus b_j}$, with the V_i and the W_j irreducible representations. If a given summand V_i meets a summand W_j non-trivially, then they must be equal because otherwise their intersection would be a non-trivial G-invariant subspace of at least one of them. The two decompositions can thus differ at most by permuting the summands. It therefore makes sense to talk about the number of times an irreducible representation V_i occurs in a representation V.

A basic tool in the representation theory of finite groups and compact Lie groups is the following.

Lemma 1.10 (Schur's lemma) *Let G be a finite group or a compact Lie group, and let V and W be finite-dimensional irreducible complex representations of G. Let $\varphi : V \to W$ be a G-linear map.*

1. *Either φ is an isomorphism or $\varphi = 0$.*
2. *If $V = W$, then there is a $\lambda \in \mathbb{C}$ such that $\varphi = \lambda \cdot I$, with I the identity map on V.*

Proof 1. Since φ is G-linear, $\ker \varphi$ is an invariant subspace of V, and since V is irreducible, this means that either $\ker \varphi = V$ (and hence $\varphi = 0$) or else $\ker \varphi = \{0\}$, so that φ is injective. In that case, $\operatorname{im} \varphi$ is a nonzero invariant subspace of W, and hence $\operatorname{im} \varphi = W$: φ is an isomorphism.

2. Since V is a complex vector space, φ must have at least one eigenvalue; i.e., there is a $\lambda \in \mathbb{C}$ such that $\ker(\varphi - \lambda I) \neq \{0\}$. But then since V is irreducible, $\ker(\varphi - \lambda I) = V$ and thus $\varphi = \lambda I$. □

Given a representation $\rho : G \to GL(V)$ of a group G, the **character** of the representation is the function

$$\chi_V(g) = \mathrm{Tr}(\rho(g)).$$

Note that if $h \in G$, then

$$\chi_V(hgh^{-1}) = \mathrm{Tr}(\rho(hgh^{-1})) = \mathrm{Tr}(\rho(h)\rho(g)\rho(h)^{-1}) = \mathrm{Tr}(\rho(g)),$$

and so χ_V is a class function on G. The following properties are easy to check.

Proposition 1.11 *Let V and W be representations of G. Then*

- $\chi_V(e) = \dim(V)$
- $\chi_{V\oplus W} = \chi_V + \chi_W$
- $\chi_{V\otimes W} = \chi_V \chi_W$
- $\chi_{V^*} = \overline{\chi_V}$.

Proof Exercise. □

Since all finite-dimensional representations can be decomposed into irreducible representations, the second property above says that all characters can be written as sums of characters corresponding to irreducible representations; these are referred to as the irreducible characters of the group. The irreducible characters satisfy two important orthogonality relations with respect to the inner product $(\cdot,\cdot)_G$ given by

$$(\alpha,\beta)_G = \begin{cases} \frac{1}{|G|}\sum_{g\in G}\alpha(g)\overline{\beta(g)}, & G \text{ finite;} \\ \int_G \alpha(g)\overline{\beta(g)}dg, & G \text{ a compact Lie group,} \end{cases}$$

where $\alpha, \beta : G \to \mathbb{C}$ are class functions. The first is the following.

Proposition 1.12 (First orthogonality relation) *The irreducible characters of a finite group or compact Lie group G are orthonormal with respect to $(\cdot,\cdot)_G$.*

In fact, one can prove that the irreducible characters form an orthonormal basis of the space of class functions on G (in the compact Lie group case, this should be interpreted as saying that the irreducible characters form a complete orthonormal system in $L^2(G)$). In the finite group case, this has the following consequence.

Proposition 1.13 (Second orthogonality relation) *Let $\chi_1, \ldots, \chi_N$ be the irreducible characters of the finite group G. Then*

$$\sum_{j=1}^{N} \chi_j(g)\overline{\chi_j(g')} = \begin{cases} \frac{|G|}{c(g)}, & g \sim g'; \\ 0, & \text{otherwise,} \end{cases}$$

where $c(g)$ is the size of the conjugacy class of g and $g \sim g'$ means that g and g' are conjugate.

Proof Let $g \in G$, and let

$$\mathbb{1}_{[g]}(h) = \begin{cases} 1, & h \sim g; \\ 0, & \text{otherwise.} \end{cases}$$

Since $\mathbb{1}_{[g]}$ is a class function and the irreducible characters are an orthonormal basis, $\mathbb{1}_{[g]}$ can be expanded as

$$\mathbb{1}_{[g]} = \sum_{j=1}^{N} (\mathbb{1}_{[g]}, \chi_j)_G \chi_j = \sum_{j=1}^{N} \left(\frac{1}{|G|} \sum_{h \in G} \mathbb{1}_{[g]}(h) \overline{\chi_j(h)} \right) \chi_j = \frac{c(g)}{|G|} \sum_{j=1}^{N} \overline{\chi_j(g)} \chi_j.$$

Multiplying both sides by $\frac{|G|}{c(g)}$ and evaluating at $g' \in G$ gives the claimed orthogonality. □

Exercise 1.14 Give a second proof by observing that the matrix $\left[\sqrt{\frac{c(g)}{|G|}} \chi_j(g) \right]$, with rows indexed by $j \in \{1, \ldots, N\}$ and columns indexed by conjugacy classes of G, is unitary.

There are many important consequences of the orthogonality relations, too many to go into here. It is worth mentioning, however, that the first orthogonality relation implies that a representation is uniquely determined by its character. Indeed, given a representation V, we have seen that it is possible to write $V = \oplus_{i=1}^{k} V_i^{\oplus a_i}$ for irreducible representations V_i and integers a_i. Then $\chi_V = \sum_{i=1}^{k} a_i \chi_{V_i}$. But since the χ_{V_i} are orthonormal, we have that $a_i = (\chi_V, \chi_{V_i})_G$. That is, the decomposition $V = \oplus_{i=1}^{k} V_i^{\oplus a_i}$ can be recovered from χ_V.

We now turn to the irreducible characters of the classical compact groups. First, we will need to introduce some basic notions about integer partitions. Given a nonnegative integer N, a **partition** λ of N is an ordered tuple $\lambda = (\lambda_1, \ldots, \lambda_k)$ with $\lambda_i \in \mathbb{N}$ for each i, $\lambda_1 \geq \lambda_2 \geq \cdots \geq \lambda_k$, and $\lambda_1 + \cdots + \lambda_k = N$. The λ_i are called the **parts** of λ. It is sometimes convenient to choose k to be larger than what might seem to be the obvious choice by tacking 0's onto the end of λ; if two partitions differ only in this final string of 0's, they are considered to be the same. The number of nonzero parts of a partition λ is called the **length** of λ and is denoted $\ell(\lambda)$. The integer N, i.e., the sum of the parts of λ, is called the **weight** of λ and is denoted $|\lambda|$.

The following altenative notation for integer partitions is often useful. Given $k \in \mathbb{N}$, we write $\lambda = (1^{a_1}, 2^{a_2}, \ldots, k^{a_k})$ for the partition of $N = a_1 + 2a_2 + \cdots + ka_k$ which has a_1 parts of size 1, a_2 parts of size 2, and so on. In this notation, the partition $(3, 3, 2, 1)$ of 9 would thus be written $(1^1, 2^1, 3^2)$.

Integer partitions are often represented by **Young diagrams**: for a partition $\lambda = (\lambda_1, \ldots, \lambda_k)$, its Young diagram is a collection of boxes drawn from the top-left corner[1], with λ_1 boxes in the top row, λ_2 boxes in the next row, and so on. The **conjugate partition** λ' of λ is then the one corresponding to the reflection of the Young diagram of λ across the diagonal. Here are the Young diagrams of the partitions $\lambda = (5,4,1)$ and $\lambda' = (3,2,2,2,1)$:

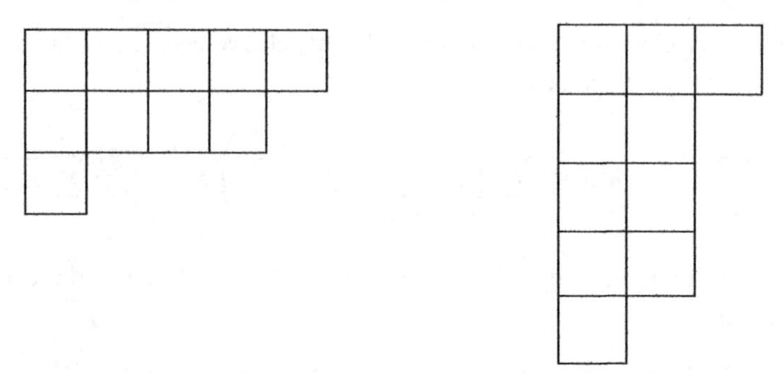

Now, for a multi-index $\alpha = (\alpha_1, \ldots, \alpha_n)$, define the anti-symmetric polynomial

$$a_\alpha(x_1, \ldots, x_n) = \sum_{\pi \in S_n} \mathrm{sgn}(\pi) x_{\pi(1)}^{\alpha_1} \cdots x_{\pi(n)}^{\alpha_n}.$$

Note that if $\sigma \in S_n$, then

$$a_\alpha(x_{\sigma(1)}, \ldots, x_{\sigma(n)}) = \mathrm{sgn}(\sigma) a_\alpha(x_1, \ldots, x_n).$$

In particular, this shows that $a_\alpha(x_1, \ldots, x_n) = 0$ if $x_i = x_j$ for any $i \neq j$ and that if any $\alpha_i = \alpha_j$ with $i \neq j$, then $a_\alpha \equiv 0$.

Assume, then, that $\alpha_1 > \alpha_2 > \cdots > \alpha_n \geq 0$; in particular, $\alpha_1 \geq n-1$, $\alpha_2 \geq n-2$, and so on. Write $\alpha = \lambda + \delta$, where $\delta = (n-1, n-2, \ldots, 0)$ and λ is a partition of length at most n. Then

$$a_\alpha(x_1, \ldots, x_n) = \sum_{\pi \in S_n} \mathrm{sgn}(\pi) \prod_{j=1}^{n} x_{\pi(j)}^{\lambda_j - n + j} = \det\left(\left[x_i^{\lambda_j + n - j}\right]_{1 \leq i,j \leq n}\right).$$

Since we have already observed that $a_\alpha(x_1, \ldots, x_n)$ vanishes if $x_i = x_j$, $a_\alpha(x_1, \ldots, x_n)$ is divisible by $x_i - x_j$ for each $i < j$, and therefore by their product, which is the Vandermonde determinant

$$\prod_{1 \leq i < j \leq n} (x_i - x_j) = \det\left(\left[x_i^{n-j}\right]_{1 \leq i,j \leq n}\right) = a_\delta(x_1, \ldots, x_n).$$

The quotient

$$s_\lambda(x_1, \ldots, x_n) = \frac{a_{\lambda+\delta}(x_1, \ldots, x_n)}{a_\delta(x_1, \ldots, x_n)}$$

[1] Some authors start from the bottom-right instead.

is thus a symmetric polynomial in $x_1, \ldots, x_n$. The polynomial s_λ is called the **Schur function** corresponding to λ and is defined as long as $\ell(\lambda) \leq n$; by convention, $s_\lambda = 0$ if $\ell(\lambda) > n$. The Schur functions $\{s_\lambda(x_1, \ldots, x_n) : \ell(\lambda) \leq n\}$ form one of many possible bases of the symmetric polynomials in $x_1, \ldots, x_n$ over $\mathbb{Z}$. Moreover, we have the following.

Theorem 1.15 *For $U \in \mathbb{U}(n)$, denote the eigenvalues of U by $e^{i\theta_1}, \ldots, e^{i\theta_n}$. The functions*

$$\chi_\lambda(U) = s_\lambda(e^{i\theta_1}, \ldots, e^{i\theta_n})$$

for $\ell(\lambda) \leq n$ are (distinct) irreducible characters of $\mathbb{U}(n)$.

These characters do not comprise a complete list of the irreducible characters of $\mathbb{U}(n)$, but in fact there are not too many more. For each $\lambda = (\lambda_1, \ldots, \lambda_n)$, with $\lambda_1 \geq \cdots \geq \lambda_n$, define λ' by

$$\lambda' = (\lambda_1 - \lambda_n, \lambda_2 - \lambda_n, \ldots, \lambda_{n-1} - \lambda_n, 0).$$

Since the λ_i are not required to be nonnegative, λ may not be a partition, but λ' always is. The functions

$$\chi_\lambda(U) = \det(U)^{\lambda_n} s_{\lambda'}(e^{i\theta_1}, \ldots, e^{i\theta_n})$$

give the complete list of the irreducible characters of $\mathbb{U}(n)$. Note from the definition of s_λ above that if the λ in the statement of Theorem 1.15 is in fact a partition, then the two definitions given for χ_λ agree.

To give similar descriptions of (at least some of) the irreducible characters for the other groups, one needs the following analogs of the Schur functions. Given a partition λ with $\ell(\lambda) \leq n$,

$$s_\lambda^{(b)}(x_1, \ldots, x_n) := \frac{\det\left(\left[x_i^{\lambda_j+n-j+\frac{1}{2}} - x_i^{-\left(\lambda_j+n-j+\frac{1}{2}\right)}\right]_{1\leq i,j\leq n}\right)}{\det\left(\left[x_i^{n-j+\frac{1}{2}} - x_i^{-\left(n-j+\frac{1}{2}\right)}\right]_{1\leq i,j\leq n}\right)},$$

$$s_\lambda^{(c)}(x_1, \ldots, x_n) := \frac{\det\left(\left[x_i^{\lambda_j+n-j} + x_i^{-(\lambda_j+n-j)}\right]_{1\leq i,j\leq n}\right)}{\det\left(\left[x_i^{n-j} + x_i^{-(n-j)}\right]_{1\leq i,j\leq n}\right)},$$

and

$$s_\lambda^{(d_1)}(x_1,\ldots,x_n)$$
$$:= \frac{\det\left(\left[\left(x_i^{\lambda_j+n-j}+x_i^{-(\lambda_j+n-j)}\right)\mathbb{1}_{\lambda_j+n-j\neq 0}+\mathbb{1}_{\lambda_j+n-j=0}\right]_{1\le i,j\le n}\right)}{\det\left(\left[\left(x_i^{n-j}+x_i^{-(n-j)}\right)\mathbb{1}_{n-j\neq 0}+\mathbb{1}_{n-j=0}\right]_{1\le i,j\le n}\right)}.$$

If $\ell(\lambda)\le n-1$, let

$$s_\lambda^{(d_2)}(x_1,\ldots,x_{n-1}) = \frac{\det\left(\left[x_i^{\lambda_j+n-j+1}-x_i^{-(\lambda_j+n-j+1)}\right]_{1\le i,j\le n}\right)}{\det\left(\left[x_i^{n-j+1}-x_i^{-(n-j+1)}\right]_{1\le i,j\le n-1}\right)}.$$

Exercise 1.16 The functions $s_\lambda^{(b)}, s_\lambda^{(c)}$, and $s_\lambda^{(d_1)}$ are polynomials in $x_1, x_1^{-1},\ldots, x_n, x_n^{-1}$, and $s_\lambda^{(d_2)}$ is a polynomial in $x_1, x_1^{-1},\ldots,x_{n-1},x_{n-1}^{-1}$.

In the case of the remaining classical compact groups, we will restrict attention to polynomial representations. A representation $\rho : G\to GL(V)$ of a matrix group is called a **polynomial representation** if there is a basis of V such that the entries of the matrix of $\rho(g)$ are polynomials in the entries of g. (If there is one such basis, this is in fact true for all bases of V.) The following then gives the analog of Theorem 1.15 for the other groups.

Theorem 1.17

1. Let $n\in\mathbb{N}$. For $U\in\mathbb{SO}(2n+1)$, denote the eigenvalues of U by $e^{\pm i\theta_1},\ldots,e^{\pm i\theta_n},1$. The irreducible polynomial representations of $\mathbb{O}(2n+1)$ are indexed by partitions λ such that $\lambda_1'+\lambda_2'\le 2n+1$. For such a partition with $\ell(\lambda)>n$, let $\tilde{\lambda}$ be the partition defined by

$$\tilde{\lambda}_i' = \begin{cases}\lambda_i', & i>1;\\ 2n+1-\lambda_1', & i=1.\end{cases}$$

Let χ_λ denote the character of the irreducible representation of $\mathbb{O}(2n+1)$ corresponding to λ. Then

$$\chi_\lambda(U) = \begin{cases} s_\lambda^{(b)}(e^{i\theta_1},\ldots,e^{i\theta_n}), & \ell(\lambda)\le n;\\ s_{\tilde{\lambda}}^{(b)}(e^{i\theta_1},\ldots,e^{i\theta_n}), & \ell(\lambda)>n.\end{cases}$$

For $U\in\mathbb{SO}^-(2n+1)$, $-U\in\mathbb{SO}(n)$ and

$$\chi_\lambda(U) = (-1)^{|\lambda|}\chi_\lambda(-U).$$

2. *Let $n \in \mathbb{N}$. For $U \in \mathbb{SO}(2n)$, denote the eigenvalues of U by $e^{\pm i\theta_1}, \dots, e^{\pm i\theta_n}$; for $U \in \mathbb{SO}^-(2n)$, denote the eigenvalues of U by $e^{\pm i\phi_1}, \dots, e^{\pm i\phi_{n-1}}, 1, -1$. The irreducible polynomial representations of $\mathbb{O}(2n)$ are indexed by partitions λ such that $\lambda'_1 + \lambda'_2 \leq 2n$. For such a partition with $\ell(\lambda) > n$, let $\tilde{\lambda}$ be the partition defined by*

$$\tilde{\lambda}'_i = \begin{cases} \lambda'_i, & i > 1; \\ 2n - \lambda'_1, & i = 1. \end{cases}$$

Let χ_λ denote the character of the irreducible representation of $\mathbb{O}(2n)$ corresponding to λ. Then for $U \in \mathbb{SO}(2n)$,

$$\chi_\lambda(U) = \begin{cases} s_\lambda^{(d_1)}(e^{i\theta_1}, \dots, e^{i\theta_n}), & \ell(\lambda) \leq n; \\ s_{\tilde{\lambda}}^{(d_1)}(e^{i\theta_1}, \dots, e^{i\theta_n}), & \ell(\lambda) > n. \end{cases}$$

For $U \in \mathbb{SO}^-(2n)$, if $\ell(\lambda) = n$, then $\chi_\lambda(U) = 0$. Otherwise,

$$\chi_\lambda(U) = \begin{cases} s_\lambda^{(d_2)}(e^{i\phi_1}, \dots, e^{i\phi_{n-1}}), & \ell(\lambda) < n; \\ -s_{\tilde{\lambda}}^{(d_2)}(e^{i\phi_1}, \dots, e^{i\phi_{n-1}}), & \ell(\lambda) > n. \end{cases}$$

3. *Let $n \in \mathbb{N}$. For $U \in \mathbb{Sp}(2n)$, denote the eigenvalues of U by $e^{\pm i\theta_1}, \dots, e^{\pm i\theta_n}$. The irreducible polynomial representations of $\mathbb{Sp}(2n)$ are indexed by partitions λ such that $\ell(\lambda) \leq n$, and the value of the character corresponding to λ at U is*

$$\chi_\lambda(U) = s_\lambda^{(c)}(e^{i\theta_1}, \dots, e^{i\theta_n}).$$

Notes and References

We have given only the barest introduction to matrix analysis on the classical compact groups; for more, the books by Horn and Johnson [53, 54] and Bhatia [11] are invaluable references.

Of the many constructions of Haar measure presented, most are part of the folklore of the field, and it seems impossible to sort out who first wrote them down. The parametrization by Euler angles is presumably the oldest, having been used by Hurwitz in 1897; see [33] for a modern perspective. A survey of methods of generating random matrices from a more computational viewpoint is given in [81].

For the Lie group theory of the classical compact groups, Bump [16] gives a beautifully written introduction; [16] also contains some of the representation theory that appears here, and the application to the empirical spectral measure of a random unitary matrix that appears in Section 4.2 of this book. The book by Fulton and Harris [47] gives an accessible and modern treatment of representation theory in general, focusing on the case of Lie groups and Lie algebras. The book [31] by Diaconis gives an introduction to the representation theory of the symmetric group and its applications in probability that a probabilist can understand.

The character theory on the classical groups in the form presented here is hard to ferret out of the literature; Littlewood's [72] and Weyl's [105] books (from 1940 and 1939, respectively) are the standard references, but while they are both gems, they are becoming increasingly inaccessible to the modern reader (especially if her area of expertise is somewhat distant). In the case of $\mathbb{O}(2n)$ and $\mathbb{O}(2n+1)$, Littlewood and Weyl deal only with the characters indexed by partitions of length at most n; the formulae presented here for those cases can be found (written slightly differently) in section 11.9 of Littlewood. The characters corresponding to the longer partitions are related by so-called modification rules due to King [66]. Ram [91] has a bottom-line summary for evalutation of characters on $\mathbb{SO}(2n)$, $\mathbb{SO}(2n+1)$, and $\mathbb{Sp}(2n)$.

2

Distribution of the Entries

2.1 Introduction

We begin this section with a few useful and important properties of Haar measure on the classical compact groups that follow easily from translation invariance and the definitions of the groups themselves. In fact, one such property has already appeared (in Exercise 1.5): the distribution of a Haar random matrix is invariant under transposition (and conjugate transposition). The next is an obvious but important symmetry of Haar measure.

Lemma 2.1 *Let U be distributed according to Haar measure in G, where G is one of $\mathbb{O}(n)$, $\mathbb{U}(n)$, $\mathbb{SO}(n)$, $\mathbb{SU}(n)$, and $\mathbb{Sp}(2n)$. Then the entries of U are identically distributed.*

Proof To each permutation $\sigma \in S_n$, associate the matrix M_σ with entries in $\{0, 1\}$, such that $m_{ij} = 1$ if and only if $\sigma(i) = j$; then $M_\sigma \in \mathbb{O}(n)$. Multiplication on the left by M_σ permutes the rows by σ^{-1} and multiplication on the right by M_σ permutes the columns by σ. One can thus move any entry of a matrix U into, say, the top-left corner by multiplication on the right and/or left by matrices in $\mathbb{O}(n)$. By the translation invariance of Haar measure, this means that all entries have the same distribution if G is $\mathbb{O}(n)$ or $\mathbb{U}(n)$.

To complete the proof for the remaining groups, the permutation matrices may need to be slightly modified. Since for $\sigma \in S_n$, $\det(M_\sigma) = \text{sgn}(\sigma)$, the permutation matrix $M_\sigma \in \mathbb{SO}(n)$ if and only if σ is even. For $G = \mathbb{SO}(n)$ or $\mathbb{SU}(n)$, one therefore replaces one of the nonzero entries of M_σ with -1 to obtain a matrix from $\mathbb{SO}(n)$; choosing which entry intelligently (or rather, failing to make the one possible non-intelligent choice) shows that any entry of $U \in \mathbb{SO}(n)$ or $\mathbb{SU}(n)$ can be moved to the top left corner by left- and right-multiplication by matrices within the group.

For the symplectic group, it is a straightforward exercise to show that the permutation matrices that are in $\mathbb{Sp}(2n)$ are exactly those which permute the indices $\{1,\ldots,n\}$ among themselves and have $\sigma(i+n) = \sigma(i)+n$ for the rest. This allows one to move any of the first n rows to the top of a random symplectic matrix and any of the first n columns all the way to the left, without changing the distribution. For the remaining rows and columns, note that a permutation with $\sigma(i) \in \{n+1,\ldots,2n\}$ for $1 \leq i \leq n$ and $\sigma(i+n) = \sigma(i)-n$ corresponds to a kind of anti-symplectic matrix: for such σ,

$$M_\sigma J M_\sigma^T = -J.$$

However, changing the signs of the entries in either all of the first n columns (or rows) or all of the second n columns (or rows) produces a symplectic matrix, which then can be used to move rows, resp. columns, of a random symplectic matrix to the top, resp. left. □

Exercise 2.2 If U is Haar-distributed in $\mathbb{U}(n)$, the distributions of $\mathrm{Re}(U_{11})$ and $\mathrm{Im}(U_{11})$ are identical.

The symmetries above can be exploited to easily calculate moments of the matrix entries, as follows.

Example Let U be Haar distributed in G, for G as above.

1. $\mathbb{E}[u_{11}] = 0$: note that Haar measure on $\mathbb{O}(n)$ and $\mathbb{U}(n)$ is invariant under multiplication on the left by

$$\begin{bmatrix} -1 & 0 & & 0 \\ 0 & 1 & & \\ & & \ddots & \\ 0 & & & 1 \end{bmatrix}; \tag{2.1}$$

   doing so multiplies the top row (so in particular u_{11}) of U by -1 but does not change the distribution of the entries. So $u_{11} \overset{d}{=} -u_{11}$, and thus $\mathbb{E}[u_{11}] = 0$. For $G = \mathbb{SO}(n)$ or $\mathbb{SU}(n)$, just change the sign of any of the remaining ones in the matrix in (2.1); for $G = \mathbb{Sp}(2n)$, change the sign of the $(n+1, n+1)$ entry in the matrix in (2.1).
2. $\mathbb{E}|u_{11}|^2 = \frac{1}{n}$: because $U \in G$, we know that $\sum_{j=1}^n |u_{1j}|^2 = 1$, and because all the entries have the same distribution, we can write

$$\mathbb{E}|u_{11}|^2 = \frac{1}{n}\sum_{j=1}^n \mathbb{E}|u_{1j}|^2 = \frac{1}{n}\mathbb{E}\left(\sum_{j=1}^n |u_{1j}|^2\right) = \frac{1}{n}.$$

   (For $G = \mathbb{Sp}(2n)$, the n should be replaced with $2n$.)

Exercise 2.3 For $U = \left[u_{ij}\right]_{j=1}^{n}$, compute $\mathrm{Cov}\left(u_{ij}, u_{k\ell}\right)$ and $\mathrm{Cov}\left(|u_{ij}|^2, |u_{k\ell}|^2\right)$ for all i, j, k, ℓ.

Understanding the asymptotic distribution of the individual entries of Haar-distributed matrices is of course more involved than just calculating the first couple of moments, but follows from classical results. Recall that one construction of Haar measure on $\mathbb{O}(n)$ involves filling the first column with a random point on the sphere. That is, the distribution of u_{11} is exactly that of x_1, where $x = (x_1, \ldots, x_n)$ is a uniform random point of $\mathbb{S}^{n-1} \subseteq \mathbb{R}^n$. The asymptotic distribution of a single coordinate of a point on the sphere has been known for more than a hundred years; the first rigorous proof is due to Borel in 1906, but it was recognized by Maxwell and others decades earlier.

Theorem 2.4 (Borel's lemma) *Let $X = (X_1, \ldots, X_n)$ be a uniform random vector in $\mathbb{S}^{n-1} \subseteq \mathbb{R}^n$. Then*

$$\mathbb{P}\left[\sqrt{n}X_1 \leq t\right] \xrightarrow{n\to\infty} \frac{1}{\sqrt{2\pi}} \int_{-\infty}^{t} e^{-\frac{x^2}{2}}\, dx;$$

that is, $\sqrt{n}X_1$ converges weakly to a Gaussian random variable, as $n \to \infty$.

The lemma is also often referred to as the *Poincaré limit*. There are many proofs; the one given below is by the method of moments. The following proposition gives a general formula for integrating polynomials over spheres to be used below.

Proposition 2.5 *Let $P(x) = |x_1|^{\alpha_1}|x_2|^{\alpha_2}\cdots|x_n|^{\alpha_n}$. Then if X is uniformly distributed on $\sqrt{n}S^{n-1}$,*

$$\mathbb{E}\left[P(X)\right] = \frac{\Gamma(\beta_1)\cdots\Gamma(\beta_n)\Gamma(\frac{n}{2})n^{\left(\frac{1}{2}\sum\alpha_i\right)}}{\Gamma(\beta_1+\cdots+\beta_n)\pi^{n/2}},$$

where $\beta_i = \frac{1}{2}(\alpha_i + 1)$ for $1 \leq i \leq n$ and

$$\Gamma(t) = \int_0^\infty s^{t-1}e^{-s}ds = 2\int_0^\infty r^{2t-1}e^{-r^2}dr.$$

The proof is essentially a reversal of the usual trick for computing the normalizing constant of the Gaussian distribution.

Proof of Borel's lemma by moments Fix $m \in \mathbb{N}$; to prove the lemma, we need to show that if for each n, Y_n is distributed as the first coordinate of a uniform random point on $\mathbb{S}^{n-1}$, then

$$\lim_{n\to\infty} \mathbb{E}\left[\left(\sqrt{n}Y_n\right)^m\right] = \mathbb{E}\left[Z^m\right], \tag{2.2}$$

where Z is a standard Gaussian random variable. Recall that the moments of the standard Gaussian distribution are

$$\mathbb{E}[Z^m] = \begin{cases} (m-1)(m-3)(m-5)\ldots(1), & m = 2k; \\ 0, & m = 2k+1. \end{cases} \tag{2.3}$$

To prove (2.2), first note that it follows by symmetry that $\mathbb{E}[X_1^{2k+1}] = 0$ for all $k \geq 0$. Next, specializing Proposition 2.5 to $P(X) = X_1^{2k}$ gives that the even moments of X_1 are

$$\mathbb{E}[X_1^{2k}] = \frac{\Gamma\left(k+\frac{1}{2}\right)\Gamma\left(\frac{1}{2}\right)^{n-1}\Gamma\left(\frac{n}{2}\right)n^k}{\Gamma\left(k+\frac{n}{2}\right)\pi^{\frac{n}{2}}}.$$

Using the functional equation $\Gamma(t+1) = t\Gamma(t)$ and the fact that $\Gamma\left(\frac{1}{2}\right) = \sqrt{\pi}$, this simplifies to

$$\mathbb{E}[X_1^{2k}] = \frac{(2k-1)(2k-3)\ldots(1)n^k}{(n+2k-2)(n+2k-4)\ldots(n)}. \tag{2.4}$$

Equation (2.2) follows immediately. □

Corollary 2.6 *For each n, let U_n be a random orthogonal matrix. Then the sequence $\{\sqrt{n}[U_n]_{1,1}\}$ converges weakly to the standard Gaussian distribution, as $n \to \infty$.*

More recent work has made it possible to give a much more precise statement that quantifies Borel's lemma; doing so has important implications for the joint distribution of the entries of a random orthogonal matrix. We first give a brief review of various notions of distance between measures.

Metrics on Probability Measures

Let X be a metric space. The following are some of the more widely used metrics on the set of Borel probability measures on X.

1. Let μ and ν be Borel probability measures on X. The **total variation distance** between μ and ν is defined by

$$d_{TV}(\mu,\nu) := \sup_{A\subseteq X} |\mu(A) - \nu(A)|,$$

where the supremum is over Borel measurable sets. Equivalently, one can define

$$d_{TV}(\mu, \nu) := \frac{1}{2} \sup_{f:X\to\mathbb{R}} \left| \int f d\mu - \int f d\nu \right|,$$

where the supremum is over functions f that are continuous, such that $\|f\|_\infty \leq 1$. When μ and ν have densities f and g (respectively) with respect to a σ-finite measure λ on X, then

$$d_{TV}(X, Y) = \frac{1}{2} \int |f - g| d\lambda.$$

The total variation distance is a very strong metric on probability measures; in particular, a discrete distribution cannot be approximated by a continuous distribution in total variation.

Exercise 2.7

1. Prove that the first two definitions are equivalent, and are equivalent to the third when the measures have density.
   Hint: The Hahn-Jordan decomposition of the signed measure $\mu - \nu$ is useful here.
2. Prove that the total variation distance between a discrete distribution and a continuous distribution is always 1.

2. The **bounded Lipschitz distance** is defined by

$$d_{BL}(\mu, \nu) := \sup_{\|g\|_{BL}\leq 1} \left| \int g\, d\mu - \int g\, d\nu \right|,$$

where the bounded-Lipschitz norm $\|g\|_{BL}$ of $g : X \to \mathbb{R}$ is defined by

$$\|g\|_{BL} := \max\left\{ \|g\|_\infty, |g|_L \right\},$$

where $|g|_L = \sup_{x\neq y} \frac{|g(x)-g(y)|}{d(x,y)}$ is the Lipschitz constant of g. If X is a separable metric space, the bounded-Lipschitz distance is a metric for the weak topology on probability measures on X (see, e.g., [39, theorem 11.3.3]).

3. The **Kolmogorov distance** for probability measures on $\mathbb{R}$ is defined by

$$d_K(\mu, \nu) := \sup_{x\in\mathbb{R}} \left| \mu\Big((-\infty, x]\Big) - \nu\Big((-\infty, x]\Big) \right|.$$

Convergence in Kolmogorov distance is in general stronger than weak convergence, because for weak convergence, the distribution functions need only converge to the limiting distribution function at its continuity points, and the convergence is not required to be uniform.

4. The L_p **Kantorovich distance** for $p \geq 1$ is defined by

$$W_p(\mu, \nu) := \inf_{\pi} \left[\int d(x,y)^p \, d\pi(x,y) \right]^{\frac{1}{p}},$$

where the infimum is over couplings π of μ and ν; that is, probability measures π on $X \times X$ such that $\pi(A \times \mathbb{R}^n) = \mu(A)$ and $\pi(\mathbb{R}^n \times B) = \nu(B)$. The L_p Kantorovich distance is a metric for the topology of weak convergence plus convergence of moments of order p or less. It is often called the L_p Wasserstein distance, and in the case of $p = 1$, the earth-mover distance.

When $p = 1$, there is the following alternative formulation:

$$W_1(\mu, \nu) := \sup_{|f|_L \leq 1} \left| \int f \, d\mu - \int f \, d\nu \right|.$$

The fact that this is an equivalent definition of W_1 to the one given above is the Kantorovich–Rubenstein theorem. There are dual representations for W_p for $p > 1$ as well, but they are more complicated and will not come up in this book.

As a slight extension of the notation defined above, if Y and Z are random variables taking values in a metric space, $d_{TV}(Y, Z)$ is defined to be the total variation distance between the distributions of Y and Z, etc.

Quantitative Asymptotics for the Entries of Haar-Distributed Matrices

As noted above, it is a consequence of Borel's lemma that the individual entries of a random orthogonal matrix are approximately Gaussian for large matrices. Borel's lemma has been strengthened considerably, as follows.

Theorem 2.8 (Diaconis–Freedman [35]) *Let X be a uniform random point on $\sqrt{n}\mathbb{S}^{n-1}$, for $n \geq 5$, and let $1 \leq k \leq n-4$. Then if Z is a standard Gaussian random vector in $\mathbb{R}^k$,*

$$d_{TV}\big((X_1, \ldots, X_k), Z\big) \leq \frac{2(k+3)}{n-k-3}.$$

That is, not only is an individual coordinate of a random point on the sphere close to Gaussian, but in fact the joint distribution of any k coordinates is close in total variation to k i.i.d. Gaussian random variables, if $k = o(n)$. In the random matrix context, this implies that for $k = o(n)$, one can approximate any k entries from the same row or column of U by independent Gaussian random variables. This led Persi Diaconis to raised the question: How many entries of U can be simultaneously approximated by independent normal random variables?

The answer to this question of course depends on the sense of approximation. In the strongest sense, namely in total variation, the sharp answer was found independently by T. Jiang and Y. Ma [58] and by K. Stewart [97], following earlier work of Diaconis–Eaton–Lauritzen [37] and Jiang [59].

Theorem 2.9 *Let $\{U_n\}$ be a sequence of random orthogonal matrices with $U_n \in \mathbb{O}(n)$ for each n, and suppose that $p_n q_n = o(n)$. Let $U_n(p_n, q_n)$ denote the top-left $p_n \times q_n$ block of U_n, and let $Z(p_n, q_n)$ denote a $p_n \times q_n$ random matrix of i.i.d. standard normal random variables. Then*

$$\lim_{n\to\infty} d_{TV}(\sqrt{n}U_n(p_n, q_n), Z(p_n, q_n)) = 0.$$

That is, a $p_n \times q_n$ principal submatrix can be approximated in total variation by a Gaussian random matrix, as long as $p_n q_n \ll n$; in particular, this recovers the theorem of Diaconis and Freedman (without the explicit rate of convergence) when $q_n = 1$. The theorem is sharp in the sense that if $p_n \sim x\sqrt{n}$ and $q_n \sim y\sqrt{n}$ for $x, y > 0$, then $d_{TV}(\sqrt{n}U_n(p_n, q_n), Z(p_n, q_n))$ does not tend to zero.

If one relaxes the sense in which entries should be simulatenously approximable by i.i.d. Gaussian variables, one can approximate a larger collection of entries, as in the following theorem. Recall that a sequence of random variables $\{X_n\}$ tends to zero in probability (denoted $X_n \xrightarrow[n\to\infty]{\mathbb{P}} 0$) if for all $\epsilon > 0$,

$$\lim_{n\to\infty} \mathbb{P}[|X_n| > \epsilon] = 0.$$

Theorem 2.10 (Jiang [59]) *For each n, let $Y_n = \left[y_{ij}\right]_{i,j=1}^n$ be an $n \times n$ matrix of independent standard Gaussian random variables and let $\Gamma_n = \left[\gamma_{ij}\right]_{i,j=1}^n$ be the matrix obtained from Y_n by performing the Gram–Schmidt process; i.e., Γ_n is a random orthogonal matrix. Let*

$$\epsilon_n(m) = \max_{1\le i\le n, 1\le j\le m} \left|\sqrt{n}\gamma_{ij} - y_{ij}\right|.$$

Then

$$\epsilon_n(m_n) \xrightarrow[n\to\infty]{\mathbb{P}} 0$$

if and only if $m_n = o\left(\frac{n}{\log(n)}\right)$.

That is, in an "in probability" sense, as many as $o\left(\frac{n^2}{\log(n)}\right)$ entries of U can be simultaneously approximated by independent Gaussians.

Theorems 2.9 and 2.10 are the subject of Section 2.3.

2.2 The Density of a Principal Submatrix

The main result of this section, important in its own right, is also the main ingredient in the proof of Theorem 2.9 on approximating the entries of a principal submatrix of a random orthogonal matrix by i.i.d. Gaussian random variables.

Theorem 2.11 *Let U be an $n \times n$ random orthogonal matrix, and let $U_{p,q}$ denote the upper-left $p \times q$ block of U. For $q \leq p$ and $p + q < n$, the random matrix $U_{p,q}$ has density with respect to Lebesgue measure, given by*

$$g(M) = \frac{w(n-p,q)}{(2\pi)^{\frac{pq}{2}} w(n,q)} \det\left(I_q - M^T M\right)^{\frac{n-p-q-1}{2}} I_0(M^T M), \tag{2.5}$$

where $w(\cdot,\cdot)$ denotes the Wishart constant:

$$w(n,q)^{-1} = \pi^{\frac{q(q-1)}{4}} 2^{\frac{nq}{2}} \prod_{j=1}^{q} \Gamma\left(\frac{n-j+1}{2}\right),$$

and $I_0(A)$ is the indicator that A has q eigenvalues (counted with multiplicity) in $(0,1)$.

The approach to the proof is via invariant theory. We first show that if $\Gamma_{p,q}$ has the density given in (2.5), then

$$U_{p,q}^T U_{p,q} \stackrel{d}{=} \Gamma_{p,q}^T \Gamma_{p,q}. \tag{2.6}$$

We then use the fact that $M \mapsto M^T M$ is a maximal invariant (to be defined below) under the action of $\mathbb{O}(p)$ to show that (2.6) implies that $U_{p,q} \stackrel{d}{=} \Gamma_{p,q}$.

Carrying out this approach requires some background on some of the classical random matrix ensembles.

Let $q \leq n$ and let Z be an $n \times q$ random matrix with entries distributed as independent standard Gaussian random variables. The $q \times q$ random matrix $S := Z^T Z$ is called a **Wishart matrix** with n degrees of freedom. The matrix S is positive definite with probability one, and its density (with respect to Lebesgue measure) on the set S_q^+ of $q \times q$ real positive definite matrices is given by

$$p(M) = w(n,q) \det(M)^{\frac{n-q-1}{2}} e^{-\frac{1}{2}\mathrm{Tr}(M)}.$$

Now let $q \leq \min\{n_1, n_2\}$, and let S_1 and S_2 be $q \times q$ Wishart matrices, with n_1 and n_2 degrees of freedom, respectively. Then $S_1 + S_2$ has rank q with probability one, and

$$B := (S_1 + S_2)^{-1/2} S_1 (S_1 + S_2)^{-1/2} \tag{2.7}$$

is said to have the **matrix-variate beta distribution** $B(n_1, n_2; I_q)$.

Note in particular that if S_1 and S_2 are as in the definition above, then since

$$I_q - (S_1 + S_2)^{-1/2} S_1 (S_1 + S_2)^{-1/2} = (S_1 + S_2)^{-1/2} S_2 (S_1 + S_2)^{-1/2},$$

B has all its eigenvalues in $(0, 1)$. The density (with respect to Lebesgue measure) of B on the set $\mathcal{P}_q$ of $q \times q$ symmetric matrices with eigenvalues in $(0, 1)$, is given by

$$g(M) = \frac{w(n_1, q) w(n_2, q)}{w(n_1 + n_2, q)} \det(M)^{(n_1 - q - 1)/2} \det(I_q - M)^{(n_2 - q - 1)/2}. \tag{2.8}$$

Lemma 2.12 *Let $q \leq p$ and $p + q < n$, and let $U_{p,q}$ be the upper-left $p \times q$ block of a Haar-distributed $U \in \mathbb{O}(n)$. Then $\Sigma_{p,q} := U_{p,q}^T U_{p,q} \in M_q(\mathbb{R})$ has the matrix-variate beta distribution $B(p, n - p; I_q)$.*

Proof It follows from the column-by-column construction of Haar measure that the first q columns of U form a random element in

$$F_{q,n} := \left\{ \begin{bmatrix} | & & | \\ U_1 & \cdots & U_q \\ | & & | \end{bmatrix} : \langle U_i, U_j \rangle = \delta_{ij} \right\},$$

whose distribution is invariant under the action of $\mathbb{O}(n)$ by multiplication on the left. The random matrix $U_{p,q}$ in the statement of the lemma is then the first p rows of a random element in $F_{q,n}$.

Now, the second Gaussian construction of Haar measure given in Section 1.2 can be generalized to produce a translation-invariant random element of $F_{q,n}$ from a collection of Gaussian random variables, as follows. Let X be an $n \times q$ random matrix with i.i.d. standard Gaussian entries, and define

$$\widetilde{U} := X(X^T X)^{-1/2}.$$

The matrix X has rank q with probability one, and so $\widetilde{U}$ is well defined, and its distribution is easily seen to be invariant under the action of $\mathbb{O}(n)$ by left-multiplication. It therefore suffices to show that the distribution of the first p rows of $\widetilde{U}^T \widetilde{U}$ is the same as the distribution of B in (2.7), with $n_1 = p$ and $n_2 = n - p$. To see this, decompose the matrix X as

$$X = \begin{bmatrix} Y \\ Z \end{bmatrix},$$

where Y is the first p rows; then

$$U_{p,q} \stackrel{d}{=} Y(Y^T Y + Z^T Z)^{-1/2},$$

and so

$$\Sigma_{p,q} \stackrel{d}{=} (S_1 + S_2)^{-1/2} S_1 (S_1 + S_2)^{-1/2},$$

where $S_1 := Y^T Y$ and $S_2 := Z^T Z$. The matrices S_1 and S_2 are $q \times q$ Wishart matrices, with p and $n - p$ degrees of freedom, respectively, and so

$$\Sigma_{p,q} = (S_1 + S_2)^{-1/2} S_1 (S_1 + S_2)^{-1/2}$$

has a matrix-variate beta distribution. □

We next confirm that the density of $U_{p,q}^T U_{p,q}$ identified in Lemma 2.12 is as it should be; that is, if $\Gamma_{p,q}$ has the claimed density of $U_{p,q}$, then $\Gamma_{p,q}^T \Gamma_{p,q}$ also has a matrix-variate beta distribution.

Lemma 2.13 *Suppose that $\Gamma_{p,q}$ is a $p \times q$ random matrix with density*

$$g(M) = \frac{w(n-p,q)}{(2\pi)^{\frac{pq}{2}} w(n,q)} \det\left(I_q - M^T M\right)^{\frac{n-p-q-1}{2}} I_0(M^T M) \tag{2.9}$$

with respect to Lebesgue measure, where $I_0(M^T M)$ is the indicator that all of the eigenvalues of $M^T M$ lie in $(0,1)$. Then $\Gamma_{p,q}^T \Gamma_{p,q}$ has the matrix-variate beta distibution $B(p, n-p; I_q)$.

Proof Let $\mathcal{X}$ be the set of $p \times q$ matrices M over $\mathbb{R}$ such that all of the eigenvalues of $M^T M$ lie in $(0,1)$.

The matrix $\Gamma_{p,q}^T \Gamma_{p,q}$ has density h on S_q^+ if and only if for all $f : S_q^+ \to \mathbb{R}$,

$$\int_{\mathcal{X}} f(M^T M) g(M) dM = \int_{S_q^+} f(A) h(A) dA.$$

Define $g^* : S_q^+ \to \mathbb{R}$ by

$$g^*(A) = \frac{w(n-p,q)}{(2\pi)^{\frac{pq}{2}} w(n,q)} \det(I_q - A)^{\frac{n-p-q-1}{2}}$$

so that $g(M) = g^*(M^T M)$ for $M \in \mathcal{X}$. Writing $f(A) = f_1(A)\varphi(A)$ with $\varphi(A) = (2\pi)^{-\frac{pq}{2}} \exp\left(-\frac{1}{2}\operatorname{Tr}(A)\right)$, we have that

$$\int_{\mathcal{X}} f(M^T M) g(M) dM = \int f_1(M^T M) g^*(M^T M) I_0(M^T M) \varphi(M^T M) dM, \tag{2.10}$$

where the integral is now over the space $M_{p,q}(\mathbb{R})$ all real $p \times q$ matrices.

Now, $\varphi(M^T M)$ is exactly the standard Gaussian density on $M_{p,q}(\mathbb{R})$, so the right-hand side of (2.10) is simply

$$\mathbb{E}[f_1(S) g^*(S) I_0(S)],$$

where S is a $q \times q$ Wishart matrix with p degrees of freedom. That is, for all $f_1 : S_q^+ \to \mathbb{R}$,

$$\int_{S_q^+} f_1(A) g^*(A) I_0(A) p(A) dA = \int_{S_q^+} f_1(A) \varphi(A) h(A) dA,$$

where p is the density of S. It follows that

$$\begin{aligned} h(A) &= \frac{g^*(A) I_0(A) p(A)}{\varphi(A)} \\ &= \frac{w(p,q) w(n-p,q)}{w(n,q)} \det(A)^{(p-q-1)/2} \det(I_q - A)^{(n-p-q-1)/2} I_0(A), \end{aligned}$$

which is exactly the density of the matrix-variate beta distribution $B(p, n - p; I_q)$. □

Finally, we show that $U_{p,q}^T U_{p,q} \stackrel{d}{=} \Gamma_{p,q}^T \Gamma_{p,q}$ implies that $U_{p,q} \stackrel{d}{=} \Gamma_{p,q}$. To do this, we need the following concept.

Definition Let $f : X \to Y$ and suppose that a group G acts on X. The function f is **invariant** under the action of G if

$$f(x) = f(g \cdot x) \quad \text{for all } g \in G.$$

The function f is a **maximal invariant** if whenever $f(x_1) = f(x_2)$, there is a $g \in G$ such that $x_1 = g \cdot x_2$.

Lemma 2.14 *Let $q \leq p$. The function $M \mapsto M^T M$ is a maximal invariant on $p \times q$ matrices of rank q, under the action of $\mathbb{O}(p)$ by left-multiplication.*

Proof Clearly, $M \mapsto M^T M$ is invariant under the action of $\mathbb{O}(p)$. Suppose, then, that $M_1^T M_1 = M_2^T M_2$. Then $\langle M_1^T M_1 v, w \rangle = \langle M_2^T M_2 v, w \rangle$ for all $v, w \in \mathbb{R}^q$; that is,

$$\langle M_1 v, M_1 w \rangle = \langle M_2 v, M_2 w \rangle \quad \text{for all } v, w \in \mathbb{R}^q.$$

It follows that if $(v_1, \ldots, v_k)$ are such that $(M_1 v_1, \ldots, M_1 v_k)$ is an orthonormal basis of $\{M_1 v : v \in \mathbb{R}^q\}$, then $(M_2 v_1, \ldots, M_2 v_k)$ is an orthonormal basis of $\{M_2 v : v \in \mathbb{R}^q\}$. Since M_1 has rank q, it acts as an injective map $\mathbb{R}^q \to \mathbb{R}^p$, and so there is a well-defined map $U : \{M_1 v : v \in \mathbb{R}^q\} \to \{M_2 v : v \in \mathbb{R}^q\}$ with $U(M_1 v) = M_2 v$ for all $v \in \mathbb{R}^q$. Extend U to a map $\widetilde{U}$ on all of $\mathbb{R}^p$ such that $\widetilde{U}$ sends an orthonormal basis to an orthonormal basis; then $\widetilde{U} \in \mathbb{O}(p)$ and $\widetilde{U} M_1 = M_2$. □

Observe that if $\tau : X \to Y$ is a maximal invariant under the action of G on X, and if $f : X \to Z$ is any G-invariant function, then there is a function

$f^* : \tau(X) \to Z$ such that $f(x) = f^*(\tau(x))$ for all $x \in X$. Indeed, if $y \in \tau(X)$, then $y = \tau(x)$ for some x. Since τ is a maximal invariant, if $y = \tau(x')$ also, then $x = gx'$, and so $f(x) = f(x')$ because f is G-invariant. Taking $f^*(y) = f(x)$ thus produces a well-defined function f^*.

Our interest in maximal invariants lies in the following proposition.

Proposition 2.15 *Suppose the compact group G acts measurably on S, and let $\tau : S \to S'$ be a maximal invariant. Suppose that X_1, X_2 are random variables in S, with G-invariant distributions. Suppose further that $\tau(X_1) \stackrel{d}{=} \tau(X_2)$. Then $X_1 \stackrel{d}{=} X_2$.*

Proof Let $f : S \to \mathbb{R}$ be bounded and measurable, and let g be distributed according to Haar measure on G. It follows by the translation invariance of Haar measure that the function

$$x \mapsto \mathbb{E}[f(g \cdot x)]$$

is G-invariant, so as discussed above, there is a function $f^* : S' \to \mathbb{R}$ such that $\mathbb{E}[f(g \cdot x)] = f^*(\tau(x))$ for each $x \in S$. If g is taken to be independent of X_1 and X_2, then by G-invariance and Fubini's theorem,

$$\begin{aligned} \mathbb{E}[f(X_1)] = \mathbb{E}[f(g \cdot X_1)] = \mathbb{E}[f^*(\tau(X_1))] \\ = \mathbb{E}[f^*(\tau(X_2))] = \mathbb{E}[f(g \cdot X_2)] = \mathbb{E}[f(X_2)]. \end{aligned}$$

That is, $X_1 \stackrel{d}{=} X_2$. □

Finally, if $\Gamma_{p,q}$ has the density claimed in Theorem 2.11 for $U_{p,q}$, then the distribution of $\Gamma_{p,q}$ is trivially seen to be $\mathbb{O}(p)$-invariant, and of course the distribution of $U_{p,q}$ is $\mathbb{O}(p)$-invariant. It thus follows from Lemmas 2.12, 2.13, and 2.14 and Proposition 2.15 that $U_{p,q} \stackrel{d}{=} \Gamma_{p,q}$.

2.3 How Much Is a Haar Matrix Like a Gaussian Matrix?

In Total Variation

Rather than proving the full version of Theorem 2.9, we will prove the important special case of square principal submatrices. The basic ideas of the proof are the same, but the assumption that the submatrix is square, so that $p_n = o(\sqrt{n})$, results in considerable technical simplification. Note also that there is no loss in assuming $p_n \to \infty$.

Theorem 2.16 *Let $\{U_n\}$ be a sequence of Haar-distributed random orthogonal matrices with $U_n \in \mathbb{O}(n)$ for each n, and suppose that $p_n \xrightarrow{n\to\infty} \infty$, with $p_n = o(\sqrt{n})$. Let $U_n(p_n)$ denote the top-left $p_n \times p_n$ block of U_n, and let Z_{p_n}*

be a $p_n \times p_n$ random matrix whose entries are i.i.d. standard normal random variables. Then

$$\lim_{n\to\infty} d_{TV}(\sqrt{n}U_n(p_n), Z_{p_n}) = 0.$$

The essential idea of the proof is the following. Recall that if random vectors X and Y have densities f and g (respectively) with respect to a σ-finite measure λ, then

$$d_{TV}(X,Y) = \frac{1}{2}\int |f-g|d\lambda.$$

Let $g_{n,p}$ denote the density of $\sqrt{n}U_n(p)$, and let φ_p denote the density of Z_p. Letting λ denote Lebesgue measure on $\mathrm{M}_p(\mathbb{R})$,

$$\begin{aligned} d_{TV}(\sqrt{n}U_n(p), Z_p) &= \frac{1}{2}\int_{\mathrm{M}_p(\mathbb{R})} |g_{n,p} - \varphi_p| d\lambda \\ &= \frac{1}{2}\int_{\mathrm{M}_p(\mathbb{R})} \left|\frac{g_{n,p}}{\varphi_p} - 1\right| \varphi_p d\lambda = \frac{1}{2}\mathbb{E}\left|\frac{g_{n,p}(Z_p)}{\varphi_p(Z_p)} - 1\right|. \end{aligned}$$

Showing that $d_{TV}(\sqrt{n}U_n(p_n), Z_{p_n}) \to 0$ as $n \to \infty$ is thus equivalent to showing that the random variable $\frac{g_{n,p_n}(Z_{p_n})}{\varphi_{p_n}(Z_{p_n})}$ tends to 1 in L^1.

To simplify the notation, write $p = p_n$ and $Z_{p_n} = Z$. From Theorem 2.11 and a change of variables,

$$g_{n,p}(Z) = \frac{w(n-p,p)}{(2\pi n)^{\frac{p^2}{2}} w(n,p)} \det\left(I_p - \frac{1}{n}Z^TZ\right)^{\frac{n-2p-1}{2}} I_0\left(\frac{1}{n}Z^TZ\right),$$

with

$$w(n,p) = \left\{\pi^{\frac{p(p-1)}{4}} 2^{\frac{np}{2}} \prod_{j=1}^{p} \Gamma\left(\frac{n-j+1}{2}\right)\right\}^{-1}$$

and $I_0\left(\frac{1}{n}Z^TZ\right)$ the indicator that the eigenvalues of $\frac{1}{n}Z^TZ$ lie in $(0,1)$. The density φ_p is given by $\varphi_p(Z) = \frac{1}{(2\pi)^{p^2/2}} \exp\left(-\frac{1}{2}\operatorname{Tr}(Z^TZ)\right)$.

If $0 < \lambda_1 < \cdots < \lambda_p$ are the eigenvalues of Z^TZ (which are indeed strictly positive and distinct with probability one when Z is a matrix of i.i.d. Gaussian variables), the densities above can be rewritten as

$$g_{n,p}(Z) = \frac{w(n-p,p)}{(2\pi n)^{\frac{p^2}{2}} w(n,p)} \prod_{j=1}^{p}\left(1 - \frac{\lambda_j}{n}\right)^{\frac{n-2p-1}{2}} \mathbb{1}_{(0,n)}(\lambda_p)$$

and

$$\varphi_p(Z) = \frac{1}{(2\pi)^{\frac{p^2}{2}}} \exp\left(-\frac{1}{2}\sum_{j=1}^{p}\lambda_j\right),$$

so that

$$\frac{g_{n,p}(Z)}{\varphi_p(Z)} = \frac{w(n-p,p)}{n^{\frac{p^2}{2}} w(n,p)} \exp\left\{\sum_{j=1}^{p}\left[\frac{\lambda_j}{2} + \frac{n-2p-1}{2}\log\left(1-\frac{\lambda_j}{n}\right)\right]\right\} \mathbb{1}_{(0,n)}(\lambda_p).$$

We first investigate the asymptotics of the coefficient.

Lemma 2.17 *If $p = o(\sqrt{n})$, then*

$$\frac{w(n-p,p)}{n^{\frac{p^2}{2}} w(n,p)} = \exp\left\{-\frac{p^3}{2n} + o(1)\right\}.$$

Proof First suppose that p is even. From the definition of $w(n,p)$,

$$\frac{w(n-p,p)}{n^{\frac{p^2}{2}} w(n,p)} = \left(\frac{2}{n}\right)^{\frac{p^2}{2}} \prod_{j=1}^{p} \frac{\Gamma\left(\frac{n-j+1}{2}\right)}{\Gamma\left(\frac{n-p-j+1}{2}\right)}.$$

Now,

$$\begin{aligned}\Gamma\left(\tfrac{n-j+1}{2}\right) &= \left(\tfrac{n-j-1}{2}\right)\left(\tfrac{n-j-3}{2}\right)\cdots\left(\tfrac{n-j-(p-1)}{2}\right)\Gamma\left(\tfrac{n-j-(p-1)}{2}\right)\\ &= \left(\frac{n}{2}\right)^{\frac{p}{2}}\Gamma\left(\tfrac{n-j-(p-1)}{2}\right)\prod_{\ell=1}^{\frac{p}{2}}\left(1-\tfrac{j+2\ell-1}{n}\right),\end{aligned}$$

and so

$$\frac{w(n-p,p)}{n^{\frac{p^2}{2}} w(n,p)} = \prod_{j=1}^{p}\prod_{\ell=1}^{\frac{p}{2}}\left(1-\tfrac{j+2\ell-1}{n}\right) = \exp\left\{\sum_{j=0}^{p-1}\sum_{\ell=1}^{\frac{p}{2}}\log\left(1-\tfrac{j+2\ell}{n}\right)\right\}.$$

For n large enough, $\frac{j+2\ell}{n} \le \frac{1}{2}$, and so

$$\left|\log\left(1-\tfrac{j+2\ell}{n}\right) + \tfrac{j+2\ell}{n}\right| \le 2\left(\tfrac{j+2\ell}{n}\right)^2,$$

and

$$\sum_{j=0}^{p-1}\sum_{\ell=1}^{\frac{p}{2}}\left(\tfrac{j+2\ell}{n}\right)^2 = \frac{1}{n^2}\left[\frac{7}{12}p^4 + \frac{1}{2}p^3 - \frac{1}{12}p^2\right] = o(1),$$

since $p = o(\sqrt{n})$.

That is,

$$\frac{w(n-p,p)}{n^{\frac{p^2}{2}}w(n,p)} = \exp\left\{-\sum_{j=0}^{p-1}\sum_{\ell=1}^{\frac{p}{2}}\frac{j+2\ell}{n} + o(1)\right\} = \exp\left\{-\frac{p^3}{2n} + o(1)\right\}.$$

When p is odd, the proof is essentially the same but requires a small tweak; the neat cancellation in the ratio of gamma functions above required p to be even. It follows from Stirling's formula, though, that $\Gamma\left(\frac{n+1}{2}\right) = \sqrt{\frac{n}{2}}\Gamma\left(\frac{n}{2}\right)\left(1 + O\left(\frac{1}{n}\right)\right)$, so that when p is odd,

$$\prod_{j=1}^{p}\frac{\Gamma\left(\frac{n-j+1}{2}\right)}{\Gamma\left(\frac{n-p-j+1}{2}\right)} = \prod_{j=1}^{p}\frac{\sqrt{\frac{n-j}{2}}\Gamma\left(\frac{n-j}{2}\right)\left(1 + O\left(\frac{1}{n}\right)\right)}{\Gamma\left(\frac{n-p-j+1}{2}\right)}.$$

The proof now proceeds along the same lines as before. □

The bulk of the proof is of course to analyze the random variable

$$L_n := \exp\left\{\sum_{j=1}^{p}\left[\frac{\lambda_j}{2} + \frac{n-2p-1}{2}\log\left(1 - \frac{\lambda_j}{n}\right)\right]\right\}\mathbb{1}_{(0,n)}(\lambda_p).$$

Proposition 2.18 *For L_n defined as above, $e^{\frac{-p^3}{2n}}L_n$ converges to 1 in probability, as n tends to infinity.*

Proof We will in fact prove the equivalent statement that $-\frac{p^3}{2n} + \log(L_n)$ tends to zero in probability.

For $x \in (0,n)$, let $f(x) = \frac{x}{2} + \frac{n-2p-1}{2}\log\left(1 - \frac{x}{n}\right)$. Then by Taylor's theorem, there is some $\xi \in (0,x)$ such that

$$f(x) = \left(\frac{2p+1}{2n}\right)x - \left(\frac{n-2p-1}{4n^2}\right)x^2 - \left(\frac{n-2p-1}{6(n-\xi)^3}\right)x^3.$$

Now, it is known that the largest eigenvalue of Z^TZ is of order p with high probability; formally, for $t \geq 0$,

$$\mathbb{P}\left[\lambda_p \geq p\left(1 + \sqrt{\frac{p}{n}} + t\right)^2\right] \leq e^{-\frac{pt^2}{2}}$$

(see, e.g., [28].) Let $\Omega_p := \left\{Z : \lambda_p \geq p\left(2 + \sqrt{\frac{p}{n}}\right)^2\right\}$; for $Z \in \Omega_p^c$, and $\xi \in (0,\lambda_i)$ for some i,

$$\left(\frac{n-2p-1}{6(n-\xi)^3}\right) \leq \left(\frac{n-2p-1}{6(n-9p)^3}\right) \leq \frac{1}{n^2},$$

for n large enough. We thus have that for $Z \in \Omega_p^c$,

$$\sum_{j=1}^{p} f(\lambda_j) = \sum_{j=1}^{p} \left[\left(\tfrac{2p+1}{2n}\right) \lambda_j - \left(\tfrac{n-2p-1}{4n^2}\right) \lambda_j^2 - \left(\tfrac{n-2p-1}{6(n-\xi(\lambda_j))^3}\right) \lambda_j^3 \right]$$
$$= \left(\tfrac{2p+1}{2n}\right) \mathrm{Tr}(Z^T Z) - \left(\tfrac{n-2p-1}{4n^2}\right) \mathrm{Tr}((Z^T Z)^2) + E\, \mathrm{Tr}((Z^T Z)^3),$$

where the random variable E has $0 \le E \le \frac{1}{n^2}$.

Now, the means and variances of $\mathrm{Tr}((Z^T Z))^k)$ are known; see, e.g., [6]. In particular, for Z a $p \times p$ matrix of i.i.d. standard Gaussian random variables,

$$\mathbb{E}\, \mathrm{Tr}((Z^T Z))^k) = \frac{p^{k+1}}{k} \binom{2k}{k+1} + O(p^k)$$

and

$$\mathrm{Var}\left[\mathrm{Tr}((Z^T Z))^k) \right] = O(p^{2k}),$$

as p tends to infinity. In particular,

$$-\frac{p^3}{2n} + \sum_{j=1}^{p} f(\lambda_j) = \left(\tfrac{2p+1}{2n}\right) \left\{ \mathrm{Tr}(Z^T Z) - \mathbb{E}\, \mathrm{Tr}(Z^T Z) \right\}$$
$$- \left(\tfrac{n-2p-1}{4n^2}\right) \left\{ \mathrm{Tr}((Z^T Z)^2) - \mathbb{E}\, \mathrm{Tr}((Z^T Z)^2) \right\}$$
$$+ E\, \mathrm{Tr}((Z^T Z)^3) + \frac{p^2}{2n} + \frac{p^3(p+1)}{n^2}.$$

By Chebychev's inequality,

$$\mathbb{P}\left[\left(\frac{2p+1}{2n}\right) \left| \mathrm{Tr}(Z^T Z) - \mathbb{E}\, \mathrm{Tr}(Z^T Z) \right| > \epsilon \right]$$
$$\le \frac{(2p+1)^2\, \mathrm{Var}\left(\mathrm{Tr}(Z^T Z)\right)}{4n^2 \epsilon^2} = O\left(\frac{p^4}{n^2}\right) = o(1).$$

Similarly,

$$\mathbb{P}\left[\left(\frac{n-2p-1}{4n^2}\right) \left| \mathrm{Tr}((Z^T Z)^2) - \mathbb{E}\, \mathrm{Tr}((Z^T Z)^2) \right| > \epsilon \right]$$
$$\le \frac{(n-2p-1)^2\, \mathrm{Var}\left(\mathrm{Tr}((Z^T Z)^2)\right)}{16 n^4 \epsilon^2} = O\left(\frac{p^4}{n^2}\right) = o(1),$$

and

$$\mathbb{P}\left[|E\,\mathrm{Tr}((Z^TZ)^3)| > \epsilon\right]$$
$$\leq \mathbb{P}\left[|\,\mathrm{Tr}((Z^TZ)^3)| > n^2\epsilon\right] \leq \frac{\mathrm{Var}\left(\mathrm{Tr}((Z^TZ)^3)\right)}{n^4\epsilon^2 - (\mathbb{E}\,\mathrm{Tr}((Z^TZ)^3))^2} = O\left(\frac{p^6}{n^4}\right) = o(1).$$

It follows that

$$\begin{aligned}
&\mathbb{P}\left[\left|-\frac{p^3}{2n} + \sum_{j=1}^{p} f(\lambda_j)\right| > \epsilon\right] \\
&\quad \leq \mathbb{P}[\Omega_p] + \mathbb{P}\left[\left(\tfrac{2p+1}{2n}\right)\left\{\mathrm{Tr}(Z^TZ) - \mathbb{E}\,\mathrm{Tr}(Z^TZ)\right\} > \frac{\epsilon}{3}\right] \\
&\quad + \mathbb{P}\left[\left(\tfrac{n-2p-1}{4n^2}\right)\left\{\mathrm{Tr}((Z^TZ)^2) - \mathbb{E}\,\mathrm{Tr}((Z^TZ)^2)\right\} > \frac{\epsilon}{3}\right] \\
&\quad + \mathbb{P}\left[|E\,\mathrm{Tr}((Z^TZ)^3)| > \frac{\epsilon}{3}\right],
\end{aligned}$$

which tends to zero as $n \to \infty$. □

To summarize: $d_{TV}(\sqrt{n}U_n(p), Z_p) \to 0$ as $n \to \infty$ if and only if the random variable $R_n := \frac{g_{n,p}(Z)}{\varphi_p(Z)}$ tends to 1 in L^1; we have shown that R_n tends to 1 in probability. Note that in fact $\mathbb{E}R_n = \int \frac{g_{n,p}(z)}{\varphi_p(z)}\varphi_p(z)dz = 1$ for every n.

Now,

$$\begin{aligned}
\mathbb{E}|R_n - 1| &= \mathbb{E}\left[|R_n - 1|\mathbb{1}_{|R_n-1|\geq\delta} + |R_n - 1|\mathbb{1}_{|R_n-1|<\delta}\right] \\
&\leq \delta + \mathbb{E}\left[|R_n - 1|\mathbb{1}_{|R_n-1|\geq\delta}\right] \\
&\leq \delta + \mathbb{E}\left[(R_n + 1)\mathbb{1}_{|R_n-1|\geq\delta}\right],
\end{aligned}$$

and so it suffices to show that for $\delta > 0$ fixed, $\mathbb{E}\left[R_n\mathbb{1}_{|R_n-1|\geq\delta}\right] \to 0$ as $n \to \infty$.

Suppose not; i.e., that there is a subsequence R_{n_k} such that $\mathbb{E}\left[R_{n_k}\mathbb{1}_{|R_{n_k}-1|\geq\delta}\right] \geq \epsilon > 0$ for all k. Since R_{n_k} does converge to 1 in probability, there is a further subsequence $R_{n_{k(i)}}$ which converges to 1 almost surely. But since $\mathbb{E}R_{n_{k(i)}} = 1$,

$$\mathbb{E}\left[R_{n_{k(i)}}\mathbb{1}_{|R_{n_{k(i)}}-1|\geq\delta}\right] = 1 - \mathbb{E}\left[R_{n_{k(i)}}\mathbb{1}_{1-\delta<R_{n_{k(i)}}<1+\delta}\right],$$

which tends to 0 by the dominated convergence theorem, in contradiction to the assumption. We may thus conclude that $\mathbb{E}|R_n - 1| \to 0$ as $n \to \infty$, completing the proof of Theorem 2.16.

In Probability

As discussed in Section 2.1, if the notion of approximation is relaxed from the very strong total variation distance all the way to an in-probability type of approximation, it is possible to approximate many more entries of a random orthogonal matrix by i.i.d. Gaussian random variables. The basic idea is to exploit the Gauss–Gram–Schmidt construction of Haar measure described in Chapter 1, and to show that, in the sense of Theorem 2.10, the Gram–Schmidt process does not change the distribution of the entries of the Gaussian random matrix very much. This is intuitively quite reasonable: when performing the Gram–Schmidt process on the $k+1$st column of the random matrix, the first step is to subtract the projection of that column onto the span of the first k columns. The original column is a Gaussian random vector whose length is typically about $\sqrt{n}$, and whose projection onto a k-dimensional subspace typically has length about $\sqrt{k}$, so that if k is not too close to n, the subtraction makes little difference. The next step is to normalize the column; since the length of a Gaussian vector is typically quite close to its mean, this normalization should not be too different from just dividing by the deterministic quantity $\sqrt{n}$.

In this section, we will give the proof of the "if" part of Theorem 2.10 only; that is, we will show that it is possible to approximate the entries of the first $o\left(\frac{n}{\log(n)}\right)$ columns of a random orthogonal matrix, in probability, by independent Gaussians.

Recall the setting of Theorem 2.10: $Y_n = [y_{ij}]_{i,j=1}^n$ is a matrix of i.i.d. standard Gaussian random variables and $\Gamma_n = [\gamma_{ij}]_{i,j=1}^n$ is the matrix obtained by performing the Gram–Schmidt process on Y_n. The random variable $\epsilon_n(m)$ is defined by

$$\epsilon_n(m) = \max_{\substack{1\le i\le n \\ 1\le j\le m}} \left|\sqrt{n}\gamma_{ij} - y_{ij}\right|,$$

and Theorem 2.10 is the statement that $\epsilon_n(m)$ tends to zero in probability if and only if $m = o\left(\frac{n}{\log(n)}\right)$.

The bulk of the proof of Theorem 2.10 is contained in the following tail inequality for $\epsilon_n(m)$.

Proposition 2.19 *Suppose that $r \in \left(0, \frac{1}{4}\right)$, $s, t > 0$, and $m \le \frac{nr}{2}$. Then*

$$\mathbb{P}\left[\epsilon_n(m_n) \ge r(s+t)+t\right] \le 2me^{-\frac{nr^2}{2}} + \frac{nm}{s}\sqrt{\frac{2}{\pi}}e^{-\frac{s^2}{2}} + mne^{\pi}\left(\frac{e}{2}\right)^{-\frac{m}{2}} + mne^{\pi}e^{-\frac{nt^2}{8m}}.$$

Assuming the Proposition, let $t > 0$ be fixed and take

$$r = \frac{1}{\log(n)} \qquad s = (\log(n))^{3/4} \qquad \tilde{m}_n = \left\lceil \frac{\delta n}{\log(n)} \right\rceil,$$

with $\delta = \min\left\{1, \frac{t^2}{24}\right\}$. Then for n large enough, $r(s+t)+t < 2t$, and so for any $m_n = o\left(\frac{n}{\log(n)}\right)$,

$$\begin{aligned}\mathbb{P}\left[\epsilon_n(m_n) \geq 2t\right] &\leq \mathbb{P}\left[\epsilon_n(\tilde{m}_n) \geq 2t\right] \\ &\leq 4\left(\frac{n}{\log(n)}\right) e^{-\frac{n}{2(\log(n))^2}} + \frac{2n^2}{\log(n)} e^{-\frac{(\log(n))^{\frac{3}{2}}}{2}} \\ &\quad + \frac{2e^{\pi} n^2}{\log(n)}\left(\frac{e}{2}\right)^{-\frac{\delta n}{2\log(n)}} + \frac{e^{\pi} n^2}{\log(n)} e^{-3\log(n)},\end{aligned}$$

which tends to zero as n tends to infinity.

Proof of Proposition 2.19 We begin by introducing a bit more notation. Let $\mathbf{y}_j$ denote the jth column of Y_n, and let $\boldsymbol{\gamma}_j$ denote the jth column of Γ_n. Given $\boldsymbol{\gamma}_1, \dots, \boldsymbol{\gamma}_{j-1}$, let

$$\mathbf{w}_j := \mathbf{y}_j - \Delta_j \qquad\qquad \Delta_j := \sum_{k=1}^{j-1} \boldsymbol{\gamma}_k \boldsymbol{\gamma}_k^T \mathbf{y}_j,$$

so that $\boldsymbol{\gamma}_j = \frac{\mathbf{w}_j}{\|\mathbf{w}_j\|}$. For convenience, take $\Delta_1 = 0$. Finally, let

$$L_j := \left|\frac{\sqrt{n}}{\|\mathbf{w}_j\|} - 1\right|.$$

Now observe that

$$\begin{aligned}\epsilon_n(m) &= \max_{1\leq j\leq m} \left\|\sqrt{n}\boldsymbol{\gamma}_j - \mathbf{y}_j\right\|_\infty \\ &= \max_{1\leq j\leq m} \left\|\frac{\sqrt{n}\mathbf{w}_j}{\|\mathbf{w}_j\|} - \mathbf{y}_j\right\|_\infty \\ &= \max_{1\leq j\leq m} \left\|\left(\frac{\sqrt{n}}{\|\mathbf{w}_j\|} - 1\right)(\mathbf{y}_j - \Delta_j) - \Delta_j\right\|_\infty \\ &\leq \left(\max_{1\leq j\leq m} L_j\right)\left(\max_{1\leq j\leq m} \left\|\mathbf{y}_j\right\|_\infty + \max_{1\leq j\leq m} \left\|\Delta_j\right\|_\infty\right) + \max_{1\leq j\leq m} \left\|\Delta_j\right\|_\infty.\end{aligned}$$

It follows that

$$\begin{aligned}&\mathbb{P}\left[\epsilon_n(m) \geq r(s+t)+t\right] \\ &\quad \leq \mathbb{P}\left[\max_{1\leq j\leq m} L_j \geq r\right] + \mathbb{P}\left[\max_{1\leq j\leq m} \left\|\mathbf{y}_j\right\|_\infty \geq s\right] + \mathbb{P}\left[\max_{1\leq j\leq m} \left\|\Delta_j\right\|_\infty \geq t\right]. \qquad (2.11)\end{aligned}$$

To estimate the first term,

$$\begin{aligned}
\mathbb{P}\left[\max_{1\le j\le m} L_j \ge r\right] &\le m \max_{1\le j\le m} \mathbb{P}\left[L_j \ge r\right] \\
&= m \max_{1\le j\le m} \mathbb{P}\left[\left|\frac{\sqrt{n}}{\|\mathbf{w}_j\|} - 1\right| \ge r\right] \\
&= m \max_{1\le j\le m} \left(\mathbb{P}\left[\frac{\sqrt{n}}{\|\mathbf{w}_j\|} \le 1 - r\right] + \mathbb{P}\left[\frac{\sqrt{n}}{\|\mathbf{w}_j\|} \ge 1 + r\right]\right).
\end{aligned}$$

On $\left(0, \frac{1}{4}\right)$, $\frac{1}{(1-r)^2} \ge 1 + 2r$, and so for each j,

$$\mathbb{P}\left[\frac{\sqrt{n}}{\|\mathbf{w}_j\|} \le 1 - r\right] = \mathbb{P}\left[\frac{\|\mathbf{w}_j\|^2}{n} \ge \frac{1}{(1-r)^2}\right] \le \mathbb{P}\left[\frac{\|\mathbf{w}_j\|^2}{n} \ge 1 + 2r\right].$$

Since $\mathbf{w}_j$ is a projection of the standard Gaussian vector $\mathbf{y}_j$, $\|\mathbf{w}_j\| \le \mathbf{y}_j$, and so

$$\begin{aligned}
\mathbb{P}\left[\frac{\|\mathbf{w}_j\|^2}{n} \ge 1 + 2r\right] \le \mathbb{P}\left[\frac{\|\mathbf{y}_j\|^2}{n} \ge 1 + 2r\right]& \\
&= \mathbb{P}\left[y_{1j}^2 + \cdots + y_{nj}^2 \ge n(1 + 2r)\right] \\
&\le e^{-\lambda n(1+2r)}\left(\mathbb{E}[e^{\lambda Z^2}]\right)^n
\end{aligned}$$

for any $\lambda > 0$, where Z is a standard Gaussian random variable. The final quantity on the right is given by

$$\left(\mathbb{E}[e^{\lambda Z^2}]\right)^n = \frac{1}{(1-2\lambda)^{\frac{n}{2}}}.$$

Taking $\lambda = \frac{1}{2} - \frac{1}{2(2r+1)}$ then gives that

$$\mathbb{P}\left[\frac{\|\mathbf{w}_j\|^2}{n} \ge 1 + 2r\right] \le \exp\left\{-n\left(r - \frac{1}{2} - \log(1 + 2r)\right)\right\} \le e^{\frac{-nr^2}{2}}$$

for $r \in \left(0, \frac{1}{4}\right)$.

Consider next

$$\mathbb{P}\left[\frac{\sqrt{n}}{\|\mathbf{w}_j\|} \ge 1 + r\right] = \mathbb{P}\left[\frac{\|\mathbf{w}_j\|^2}{n} \le \frac{1}{(1+r)^2}\right] \le \mathbb{P}\left[\frac{\|\mathbf{w}_j\|^2}{n} \le 1 - r\right],$$

since $\frac{1}{(1+r)^2} \le 1 - r$ for $r \in \left(0, \frac{1}{4}\right)$.

Recall that $\mathbf{w}_j$ is the orthogonal projection of $\mathbf{y}_j$ onto $\langle \boldsymbol{\gamma}_1, \dots \boldsymbol{\gamma}_{j-1} \rangle^{\perp}$; conditional on $(\mathbf{y}_1, \dots, \mathbf{y}_{j-1})$, $\mathbf{w}_j$ is a standard Gaussian random vector in the $(n-j+1)$-dimensional subspace $\langle \boldsymbol{\gamma}_1, \dots \boldsymbol{\gamma}_{j-1} \rangle^{\perp} \subseteq \mathbb{R}^n$. It follows that

$$\left\| \mathbf{w}_j \right\|^2 \stackrel{d}{=} Z_1^2 + \cdots + Z_{n-j+1}^2,$$

where the Z_i are i.i.d. standard Gaussian random variables. Now proceeding similarly to the argument above, for $1 \leq j \leq m$, and any $\lambda > 0$,

$$\begin{aligned}
\mathbb{P}\left[\frac{\left\| \mathbf{w}_j \right\|^2}{n} \leq 1 - r \right] &= \mathbb{P}\left[Z_1^2 + \cdots + Z_{n-j+1}^2 \leq n(1-r) \right] \\
&\leq \mathbb{P}\left[Z_1^2 + \cdots + Z_{n-m}^2 \leq n(1-r) \right] \\
&\leq e^{\lambda n(1-r)} \left(\mathbb{E}[e^{-\lambda Z^2}] \right)^{n-m} \\
&= e^{\lambda n(1-r)} \left(\frac{1}{\sqrt{1+2\lambda}} \right)^{n-m}.
\end{aligned}$$

Taking $\lambda = \frac{n-m}{2n(1-r)} - \frac{1}{2}$ gives that

$$\mathbb{P}\left[\frac{\left\| \mathbf{w}_j \right\|^2}{n} \leq 1 - r \right] \leq \exp\left\{ \frac{nr}{2} - \frac{m}{2} + \frac{n-m}{2} \log\left(\frac{n-m}{n(1-r)} \right) \right\},$$

and for $r \in \left(0, \frac{1}{4}\right)$ and $m \leq \frac{nr}{2}$, this last expression is bounded by $e^{-\frac{nr^2}{2}}$ as well.

Together then, we have that

$$\mathbb{P}\left[\max_{1 \leq j \leq m} L_j \geq r \right] \leq 2me^{-\frac{nr^2}{2}}. \tag{2.12}$$

The second term in (2.11) is almost trivial: since the $\mathbf{y}_j$ are i.i.d. Gaussian vectors in $\mathbb{R}^n$, if Z is a standard Gaussian random variable, then

$$\mathbb{P}\left[\max_{1 \leq j \leq m} \left\| \mathbf{y}_j \right\|_\infty \geq s \right] \leq nm\mathbb{P}\left[|Z| \geq s \right] \leq \frac{nm}{s} \sqrt{\frac{2}{\pi}} e^{-\frac{s^2}{2}}. \tag{2.13}$$

Finally, consider the random variable $\Delta_j = P_{j-1}\mathbf{y}_j$, where $P_{j-1} = \sum_{k=1}^{j-1} \boldsymbol{\gamma}_k \boldsymbol{\gamma}_k^T$ is the matrix of orthogonal projection onto $\langle \boldsymbol{\gamma}_1, \dots, \boldsymbol{\gamma}_{j-1} \rangle$. Since P_{j-1} depends only on $\mathbf{y}_1, \dots, \mathbf{y}_{j-1}$, P_{j-1} and $\mathbf{y}_j$ are independent. Moreover, by the Gauss–Gram–Schmidt construction of Haar measure, $(\boldsymbol{\gamma}_1, \dots, \boldsymbol{\gamma}_{j-1})$ are distributed as the first $j-1$ columns of a Haar-distributed random matrix, and so

$$P_{j-1} \stackrel{d}{=} U(I_{j-1} \oplus O_{n-j+1})U^T,$$

where U is distributed according to Haar measure on $\mathbb{O}(n)$ and is independent of $\mathbf{y}_j$. Using the independence together with the rotation invariance of the distribution of $\mathbf{y}_j$, it follows that

$$P_{j-1}\mathbf{y}_j \stackrel{d}{=} U \begin{bmatrix} Z_1 \\ \vdots \\ Z_{j-1} \\ 0 \\ \vdots \\ 0 \end{bmatrix} =: U\mathbf{z}_j,$$

where $Z_1, \ldots, Z_{j-1}$ are i.i.d. Gaussian random variables, independent of U. Conditioning now on the Z_i, let $R_j \in \mathbb{O}(n)$ be such that $R_j\mathbf{z}_j = \sqrt{(Z_1^2 + \cdots + Z_{j-1}^2)}\mathbf{e}_1$; it follows by the rotational invariance of Haar measure that, conditional on the Z_i,

$$U\mathbf{z}_j \stackrel{d}{=} UR_j\mathbf{z}_j = (Z_1^2 + \cdots + Z_{j-1}^2)\mathbf{u}_1.$$

It thus follows that for each $j \in \{2, \ldots, m\}$,

$$\mathbb{P}\left[\|\Delta_j\|_\infty \geq t\right] = \mathbb{P}\left[\|\theta\|_\infty \geq \frac{t}{\sqrt{Z_1^2 + \cdots + Z_{j-1}^2}}\right],$$

where θ is uniformly distributed on the unit sphere $\mathbb{S}^{n-1} \subseteq \mathbb{R}^n$. Lévy's lemma (see Section 5.1) gives that for a single coordinate θ_k of a uniform random vector on $\mathbb{S}^{n-1}$,

$$\mathbb{P}[|\theta_k| > t] \leq e^{\pi - \frac{nt^2}{4}}.$$

Conditioning on the Z_i, Lévy's lemma thus gives that

$$\begin{aligned} \mathbb{P}\left[\max_{1 \leq j \leq m} \|\Delta_j\| \geq t\right] &\leq nm\mathbb{E}\left[\mathbb{P}\left[|\theta_1| \geq \frac{t}{\sqrt{Z_1^2 + \cdots + Z_m^2}} \middle| Z_1, \ldots, Z_m\right]\right] \\ &\leq nm\mathbb{E}\left[\exp\left\{\pi - \frac{nt^2}{4\sum_{k=1}^m Z_k^2}\right\}\right]. \end{aligned}$$

Estimating as above,

$$\mathbb{P}\left[\sum_{k=1}^m Z_k^2 \geq x_0\right] \leq \exp\left\{\frac{m}{2} - \frac{x_0}{2} - \frac{m}{2}\log\left(\frac{m}{x_0}\right)\right\}$$

for $x_0 > m$, and so

$$\mathbb{E}\left[\exp\left\{-\frac{nt^2}{4\sum_{k=1}^m Z_k^2}\right\}\right] \le e^{-\frac{nt^2}{4x_0}} + \exp\left\{\frac{m}{2} - \frac{x_0}{2} - \frac{m}{2}\log\left(\frac{m}{x_0}\right)\right\},$$

and choosing $x_0 = 2m$ gives that

$$nme^{\pi}\mathbb{E}\left[\exp\left\{-\frac{t^2}{4\sum_{k=1}^m Z_k^2}\right\}\right] \le nme^{\pi-\frac{nt^2}{8m}} + nm\exp\left\{\pi - \frac{m}{2}(1-\log(2))\right\}.$$

This completes the proof of the Proposition. □

2.4 Arbitrary Projections

A deficiency of Theorem 2.9 is that it applies only to entries in a principal submatrix of U. Thus one may conclude that $o(n)$ entries of U can be simultaneously approximated in total variation by i.i.d. Gaussian random variables, *if* those entries are those of a $p \times q$ principal submatrix with $pq = o(n)$; the original question assumed no such restriction, and indeed, the fact that Haar measure is invariant under multiplication by any orthogonal matrix suggests that this restriction is too strong. The following result overcomes this difficulty, but in the weaker L^1 Kantorovich metric.

Theorem 2.20 (Chatterjee–Meckes) *Let $U \in \mathbb{O}(n)$ be distributed according to Haar measure, and let $A_1, \ldots, A_k$ be $n \times n$ matrices over $\mathbb{R}$ satisfying $\mathrm{Tr}(A_i A_j^T) = n\delta_{ij}$; that is, $\left\{\frac{1}{\sqrt{n}}A_i\right\}_{1\le i\le k}$ is orthonormal with respect to the Hilbert–Schmidt inner product. Define the random vector X by*

$$X := (\mathrm{Tr}(A_1U), \mathrm{Tr}(A_2U), \ldots, \mathrm{Tr}(A_kU))$$

in $\mathbb{R}^k$, and let $Z = (Z_1, \ldots, Z_k)$ be a random vector whose components are independent standard normal random variables. Then for $n \ge 2$,

$$W_1(X,Z) \le \frac{\sqrt{3}k}{n-1}.$$

In particular, if E_{ij} denotes the matrix with 1 as the i-jth entry and zeroes elsewhere, then choosing the A_ℓ to be $\{\sqrt{n}E_{ij}\}$ for some collection of pairs (i,j) gives that *any* collection of $o(n)$ entries of U can be simultaneously approximated (in W_1) by i.i.d. Gaussians. However, the theorem is more general: it may be that all of the entries of U appear in $\mathrm{Tr}(A_iU)$, for some or all i. Indeed, the general form of the vector X above is that of a projection of a random element of $\mathbb{O}(n)$ onto a subspace of $\mathrm{M}_n(\mathbb{R})$ of rank k. A Gaussian distribution on $\mathrm{M}_n(\mathbb{R})$

has the property that all of its projections onto lower-dimensional subspaces are also Gaussian; the theorem above can thus be seen as a coordinate-free comparison between the Haar measure on $\mathbb{O}(n)$ and standard Gaussian measure on $\mathrm{M}_n(\mathbb{R})$.

The proof of Theorem 2.20 makes use of the following framework for proving multivariate central limit theorems, which is a version of Stein's method of exchangeable pairs. For a proof of the theorem and further discussion, see [18].

Theorem 2.21 *Let X be a random vector in $\mathbb{R}^k$ with $\mathbb{E}X = 0$, and for each $\epsilon > 0$, let X_ϵ be a random vector with $X \overset{d}{=} X_\epsilon$, such that*

$$\lim_{\epsilon \to 0} X_\epsilon = X$$

almost surely. Let Z be a standard normal random vector in $\mathbb{R}^k$. Suppose there are deterministic $\lambda(\epsilon)$ and $\sigma^2 > 0$, and a random matrix F such that the following conditions hold.

1. $\frac{1}{\lambda(\epsilon)}\mathbb{E}\left[(X_\epsilon - X)\middle|X\right] \xrightarrow[\epsilon \to 0]{L_1} -X.$
2. $\frac{1}{2\lambda(\epsilon)}\mathbb{E}\left[(X_\epsilon - X)(X_\epsilon - X)^T\middle|X\right] \xrightarrow[\epsilon \to 0]{L_1} \sigma^2 I_k + \mathbb{E}\left[F\middle|X\right].$
3. *For each $\rho > 0$,*

$$\lim_{\epsilon \to 0} \frac{1}{\lambda(\epsilon)}\mathbb{E}\left[\left|X_\epsilon - X\right|^2 \mathbb{1}_{\{|X_\epsilon - X|^2 > \rho\}}\right] = 0.$$

Then

$$W_1(X, \sigma Z) \leq \frac{1}{\sigma}\mathbb{E}\|F\|_{H.S.} \tag{2.14}$$

The idea of the theorem is the following. Suppose that the random vector X has "continuous symmetries" that allow one to make a small (parametrized by ϵ) random change to X, which preserves its distribution. If X were exactly Gaussian and this small random change could be made so that (X, X_ϵ) were jointly Gaussian, then we would have (for some parametrization of the size of the change) that $X_\epsilon \overset{d}{=} \sqrt{1 - \epsilon^2}X + \epsilon Y$ for X and Y independent. The conditions of the theorem are approximate versions of what happens, up to third order, in this jointly Gaussian case.

The other technical tool needed for the proof of Theorem 2.20 is the following lemma, which gives formulae for the fourth-order mixed moments of entries of a random orthogonal matrix. The proof uses the same ideas as those in Example 2.1, and is a good exercise in symmetry exploitation and tedious calculation.

Lemma 2.22 *If $U = [u_{ij}]_{i,j=1}^n$ is an orthogonal matrix distributed according to Haar measure, then $\mathbb{E}\left[\prod u_{ij}^{k_{ij}}\right]$ is nonzero if and only if the number of entries from each row and from each column is even. The fourth-degree mixed moments are as follows: for all $i, j, r, s, \alpha, \beta, \lambda, \mu$,*

$$\begin{aligned}
&\mathbb{E}\big[u_{ij}u_{rs}u_{\alpha\beta}u_{\lambda\mu}\big] \\
&= -\frac{1}{(n-1)n(n+2)}\Big[\delta_{ir}\delta_{\alpha\lambda}\delta_{j\beta}\delta_{s\mu} + \delta_{ir}\delta_{\alpha\lambda}\delta_{j\mu}\delta_{s\beta} + \delta_{i\alpha}\delta_{r\lambda}\delta_{js}\delta_{\beta\mu} \\
&\qquad + \delta_{i\alpha}\delta_{r\lambda}\delta_{j\mu}\delta_{\beta s} + \delta_{i\lambda}\delta_{r\alpha}\delta_{js}\delta_{\beta\mu} + \delta_{i\lambda}\delta_{r\alpha}\delta_{j\beta}\delta_{s\mu}\Big] \\
&\quad + \frac{n+1}{(n-1)n(n+2)}\Big[\delta_{ir}\delta_{\alpha\lambda}\delta_{js}\delta_{\beta\mu} + \delta_{i\alpha}\delta_{r\lambda}\delta_{j\beta}\delta_{s\mu} + \delta_{i\lambda}\delta_{r\alpha}\delta_{j\mu}\delta_{s\beta}\Big].
\end{aligned} \tag{2.15}$$

Proof of Theorem 2.20 We begin by constructing an exchangeable pair (U, U_ϵ) of random orthogonal matrices. Let U be a Haar-distributed element of $\mathbb{O}(n)$, and let A_ϵ be the rotation

$$A_\epsilon = \begin{bmatrix} \sqrt{1-\epsilon^2} & \epsilon \\ -\epsilon & \sqrt{1-\epsilon^2} \end{bmatrix} \oplus I_{n-2} = I_n + \begin{bmatrix} -\frac{\epsilon^2}{2} + \delta & \epsilon \\ -\epsilon & -\frac{\epsilon^2}{2} + \delta \end{bmatrix} \oplus \mathbf{0}_{n-2},$$

where $\delta = O(\epsilon^4)$. Let V be Haar-distributed in $\mathbb{O}(n)$, independent of U, and define

$$U_\epsilon = VA_\epsilon V^T U.$$

That is, U_ϵ is a translation of U within $\mathbb{O}(n)$ by a rotation of size $\arcsin(\epsilon)$ in a random two-dimensional subspace of $\mathbb{R}^k$, and in particular, it follows from the translation invariance of Haar measure that $U_\epsilon \overset{d}{=} U$. Finally, let

$$X_\epsilon = (\mathrm{Tr}(A_1 U_\epsilon), \ldots, \mathrm{Tr}(A_k U_\epsilon)).$$

Let K denote the first two columns of V and

$$C_2 = \begin{bmatrix} 0 & 1 \\ -1 & 0 \end{bmatrix}.$$

Then

$$U_\epsilon - U = \left[\left(\frac{-\epsilon^2}{2} + O(\epsilon^4)\right) KK^T + \epsilon Q\right] U, \tag{2.16}$$

where $Q = KC_2K^T$. The entries of the matrices KK^T and Q are

$$(KK^T)_{ij} = u_{i1}u_{j1} + u_{i2}u_{j2} \qquad\qquad (Q)_{ij} = u_{i1}u_{j2} - u_{i2}u_{j1}.$$

It then follows from Lemmas 2.1 and 2.22 that

$$\mathbb{E}\big[KK^T\big] = \frac{2}{n}I_n \qquad \mathbb{E}\big[Q\big] = 0,$$

thus

$$\begin{aligned}
\lim_{\epsilon\to 0}\frac{n}{\epsilon^2}\mathbb{E}\big[(X_\epsilon - X)_i\big|U\big]
&= \lim_{\epsilon\to 0}\frac{n}{\epsilon^2}\mathbb{E}\left[\mathrm{Tr}[A_i(U_\epsilon - U)]\big|U\right]\\
&= \lim_{\epsilon\to 0}\frac{n}{\epsilon^2}\left[\left(-\frac{\epsilon^2}{2}+O(\epsilon^4)\right)\mathbb{E}\left[\mathrm{Tr}(A_iKK^TU)\big|U\right]+\epsilon\mathbb{E}\left[\mathrm{Tr}(A_iQU)\big|U\right]\right]\\
&= -X_i.
\end{aligned}$$

Condition 1 of Theorem 2.21 is thus satisfied with $\lambda(\epsilon) = \frac{\epsilon^2}{n}$. Condition 3 is immediate from the fact that $\frac{|X_\epsilon - X|^2}{\lambda(\epsilon)}$ is bounded independent of ϵ and X_ϵ converges pointwise to X.

The random matrix F of condition 2 is computed as follows. For notational convenience, write $A_i = A = (a_{pq})$, $A_j = B = (b_{\alpha\beta})$, and $U = (u_{ij})$. By (2.16),

$$\begin{aligned}
\lim_{\epsilon\to 0}\frac{n}{2\epsilon^2}\mathbb{E}\Big[(X_\epsilon - X)_i(X_\epsilon - X)_j\Big|U\Big]
&= \frac{n}{2}\mathbb{E}\left[\mathrm{Tr}(AQU)\,\mathrm{Tr}(BQU)\big|U\right]\\
&= \frac{n}{2}\mathbb{E}\left[\sum_{p,q,r,\alpha,\beta,\gamma} a_{pq}b_{\alpha\beta}u_{rp}u_{\gamma\alpha}q_{qr}q_{\beta\gamma}\middle| U\right]\\
&= \frac{n}{2}\left[\sum_{p,q,r,\alpha,\beta\gamma} a_{pq}b_{\alpha\beta}u_{rp}u_{\gamma\alpha}\left(\frac{2}{n(n-1)}\right)(\delta_{q\beta}\delta_{r\gamma}-\delta_{q\gamma}\delta_{r\beta})\right]\\
&= \frac{1}{(n-1)}\left[\langle UA, UB\rangle_{H.S.} - \mathrm{Tr}(AUBU)\right]\\
&= \frac{1}{(n-1)}\left[\langle A, B\rangle_{H.S.} - \mathrm{Tr}(UAUB)\right]\\
&= \frac{1}{(n-1)}\left[n\delta_{ij} - \mathrm{Tr}(UAUB)\right].
\end{aligned} \tag{2.17}$$

Thus

$$F = \frac{1}{(n-1)}\mathbb{E}\left[\left[\delta_{ij} - \mathrm{Tr}(A_iUA_jU)\right]_{i,j=1}^k\middle|X\right].$$

Claim: If $n \geq 2$, then $\mathbb{E}\left[\mathrm{Tr}(A_iUA_jU) - \delta_{ij}\right]^2 \leq 3$ for all i and j.

With the claim, for $n \geq 2$,

$$\mathbb{E}\|F\|_{H.S.} \leq \sqrt{\mathbb{E}\|F\|_{H.S.}^2} \leq \frac{\sqrt{3}k}{n-1},$$

thus completing the proof.

To prove the claim, first observe that Lemma 2.22 implies

$$\mathbb{E}\big[\operatorname{Tr}(A_iUA_jU)\big] = \frac{1}{n}\langle A_i, A_j\rangle = \delta_{ij}.$$

Again writing $A_i = A$ and $A_j = B$, applying Lemma 2.22 yields

$$\begin{aligned}
&\mathbb{E}\Big[\operatorname{Tr}(AUBU)\Big]^2 \\
&\quad = \mathbb{E}\left[\sum_{\substack{p,q,r,s\\ \alpha,\beta,\mu,\lambda}} a_{sp}a_{\mu\alpha}b_{qr}b_{\beta\lambda}u_{pq}u_{rs}u_{\alpha\beta}u_{\lambda\mu}\right] \\
&\quad = -\frac{2}{(n-1)n(n+2)}\Big[\operatorname{Tr}(A^TAB^TB) + \operatorname{Tr}(AB^TAB^T) + \operatorname{Tr}(AA^TBB^T)\Big] \\
&\qquad + \frac{n+1}{(n-1)n(n+2)}\Big[2\,\langle A, B\rangle_{H.S.} + \|A\|_{H.S.}^2\|B\|_{H.S.}^2\Big].
\end{aligned}$$

Since the Hilbert–Schmidt norm is submultiplicative,

$$\operatorname{Tr}(AB^TAB^T) \leq \|AB^T\|_{H.S.}^2 \leq \|A\|_{H.S.}^2\|B\|_{H.S.}^2 = n^2,$$

and the other two summands of the first line are non-positive. Also,

$$2\,\langle A, B\rangle_{H.S.} + \|A\|_{H.S.}^2\|B\|_{H.S.}^2 = n^2 + 2n\delta_{ij},$$

Thus

$$\mathbb{E}\left[\operatorname{Tr}(A_iUA_jU) - \delta_{ij}\right]^2 \leq \frac{2n^2 + (n+1)(n^2+2n\delta_{ij}) - (n-1)n(n+2)\delta_{ij}}{(n-1)n(n+2)} \leq 3.$$

□

The following is the analog of Theorem 2.20 for the unitary group; the proof is essentially the same, using the unitary analog of Lemma 2.22.

Theorem 2.23 (Chatterjee–Meckes) *Let $U \in \mathbb{U}(n)$ be distributed according to Haar measure, and let $A_1, \ldots, A_k$ be $n \times n$ matrices over $\mathbb{C}$ satisfying $\operatorname{Tr}(A_iA_j^*) = n\delta_{ij}$. Define the random vector X by*

$$X := (\operatorname{Tr}(A_1U), \operatorname{Tr}(A_2U), \ldots, \operatorname{Tr}(A_kU))$$

in $\mathbb{C}^k$, and let $Z = (Z_1, \dots, Z_k)$ be a random vector whose components are independent standard complex normal random variables. There is a constant c, independent of n, such that

$$W_1(X, Z) \leq \frac{ck}{n}.$$

Remark: The constant is asymptotically given by $\sqrt{2}$; for $n \geq 4$, it can be taken to be 3.

Notes and References

The paper [35] gives an extensive history of Borel's lemma. For thorough discussions of metrics on probability measures, their relationships, and terminology, see the books by Dudley [39] or Villani [102].

Section 2.2 follows the derivation of the density of a submatrix given in Eaton [41]. For more on Wishart matrices, see Muirhead's book [85], and for notation, alternate definitions, and generalizations of the matrix-variate beta distribution, see [29].

The univariate version of Theorem 2.20, namely a central limit theorem for $\text{Tr}(AU)$ where A is a fixed matrix and U is Haar-distributed on $\mathbb{O}(n)$, was first proved by d'Aristotile–Diaconis–Newman [27] as a step in proving the following.

Theorem 2.24 (d'Aristotile–Diaconis–Newman) *Let U be a Haar-distributed matrix in $\mathbb{O}(n)$. Let $\{\beta_1, \dots, \beta_{k_n}\}$ be a subset of the entries of U, ordered lexicographically. For $\ell \in \{1, \dots, k_n\}$ and $t \in [0, 1]$, let*

$$S_\ell^{(n)} = \sqrt{\frac{n}{k_n}} \sum_{j=1}^{\ell} \beta_j \qquad X_n(t) = S_{[k_n t]}^{(n)}.$$

If $k_n \nearrow \infty$, then $X_n \Longrightarrow W$, a standard Brownian motion, as $n \to \infty$.

Other approaches to the univariate case appeared in [75] and [64]. An alternative approach to weak convergence in the multivariate case was given in [22].

The idea of using Stein's method together with infinitesimal random rotations was first used by Stein in [96] to get fast rates of convergence to a Gaussian distribution for $\text{Tr}(U^k)$, for $k \in \mathbb{N}$ fixed and U distributed according to Haar measure on $\mathbb{O}(n)$. Slightly better bounds were obtained simultaneously by Johansson [60], and so Stein's argument remained a hidden gem for several years (it can now be found online in the Stanford statistics department's repository of technical reports).

Lemma 2.22 gives a formula for computing mixed moments up to order four of entries of a random orthogonal matrix by exploiting the symmetries of Haar measure; the analog for the unitary group can be found in [18]. A systematic approach called the Weingarten calculus for computing moments of all orders was developed by B. Collins [20] in the unitary case and extended to all the classical compact groups by Collins and Śniady [21]. This approach makes heavy use of the representation theory of the classical groups, in particular exploiting Schur–Weyl duality and its analogs, and was the basis for the approach to weak convergence of $\mathbb{R}^k$-valued linear functions on the groups given in [22].

3

Eigenvalue Distributions

Exact Formulas

3.1 The Weyl Integration Formula

Suppose U is a Haar-distributed random matrix. Then U has eigenvalues, all of which lie on the unit circle $\mathbb{S}^1 \subseteq \mathbb{C}$. Since U is random, its set of eigenvalues is a *random point process*; that is, it is a collection of n random points on $\mathbb{S}^1$. The eigenvalue process of a Haar-distributed random matrix has many remarkable properties, the first of which is that there is an explicit formula (due to H. Weyl) for its density. The situation is simplest for random unitary matrices.

Theorem 3.1 (Weyl integration formula on $\mathbb{U}(n)$) *The unordered eigenvalues of an $n \times n$ random unitary matrix have eigenvalue density*

$$\frac{1}{n!\,(2\pi)^n} \prod_{1 \le j < k \le n} |e^{i\theta_j} - e^{i\theta_k}|^2,$$

with respect to $d\theta_1 \cdots d\theta_n$ on $[0, 2\pi)^n$. That is, for any $g : \mathbb{U}(n) \to \mathbb{R}$ with

$$g(U) = g(VUV^*) \qquad \textit{for any } U, V \in \mathbb{U}(n),$$

(i.e., g is a class function), if U is Haar-distributed on $\mathbb{U}(n)$, then

$$\mathbb{E}g(U) = \frac{1}{n!\,(2\pi)^n} \int_{[0,2\pi)^n} \tilde{g}(\theta_1, \ldots, \theta_n) \prod_{1 \le j < k \le n} |e^{i\theta_j} - e^{i\theta_k}|^2 d\theta_1 \cdots d\theta_n,$$

where $\tilde{g} : [0, 2\pi)^n \to \mathbb{R}$ is the (necessarily symmetric) expression of $g(U)$ as a function of the eigenvalues of U.

The proof of the Weyl integration formula makes heavy use of the Lie group structure of $\mathbb{U}(n)$. In this section, we attempt to give a reasonably accessible treatment of the integration formula, stating some background results without proof and glossing over some details; see the end-of-chapter notes for further references.

As a preliminary, we state the following Fubini-like theorem in the Lie group context. Recall that if G is a group and $H \subseteq G$ is a subgroup, then the quotient G/H is the set of all cosets gH, where $g \in G$. If G is a locally compact Lie group and H is a closed subgroup, then G/H can be endowed with the *quotient topology*, which makes G/H into a locally compact Hausdorff space. Recall also from Section 1.1 that every compact Lie group has a Haar measure; that is, a (unique) probability measure invariant under both left and right translations.

Proposition 3.2 *Let G be a compact Lie group and let H be a closed subgroup; let μ_G and μ_H denote the Haar measures on G and H, respectively. There exists a regular Borel measure $\mu_{G/H}$ on G/H that is invariant under left-translation by elements of G, and which may be normalized such that for any continuous $\varphi : G \to \mathbb{R}$,*

$$\int_G \varphi(g) d\mu_G(g) = \int_{G/H} \int_H \varphi(gh) d\mu_H(h) d\mu_{G/H}(gH). \tag{3.1}$$

Observe that we have implicitly used the fact that $gH \mapsto \int_H \varphi(gh) d\mu_H(h)$ is well defined, which follows from the translation invariance of μ_H: for any $\tilde{h} \in H$,

$$\int_H \varphi(g\tilde{h}h) d\mu_H(h) = \int_H \varphi(gh) d\mu_H(h).$$

Corollary 3.3 *Let G and H be as above, and suppose that $\varphi : G \to \mathbb{R}$ is constant on cosets; i.e., $\varphi(g_1) = \varphi(g_2)$ for all g_1, g_2 such that $g_1 = g_2 h$ for some $h \in H$. Then*

$$\int_G \varphi(g) d\mu_G(g) = \int_{G/H} \varphi(g) d\mu_{G/H}(gH),$$

where the integrand $\varphi(g)$ on the right-hand side is the common value of φ on the coset gH.

Proof For all $h \in H$, $\varphi(gh) = \varphi(g)$ since φ is constant on cosets, and so the inner integrand on the right-hand side of (3.1) is constant, and μ_H was chosen to be a probability measure. □

The central idea of the proof of the Weyl integration formula is the following.

Lemma 3.4 *Let $\mathbb{T} \subseteq \mathbb{U}(n)$ denote the diagonal elements of $\mathbb{U}(n)$, and let $\mathbb{T}' \subseteq \mathbb{T}$ denote those elements of $\mathbb{T}$ with distinct diagonal entries. Let $\mathbb{U}(n)' \subseteq \mathbb{U}(n)$ denote the $n \times n$ unitary matrices with distinct eigenvalues. Then the map*

$$\begin{aligned} \psi : (\mathbb{U}(n)/\mathbb{T}) \times \mathbb{T}' &\to \mathbb{U}(n)' \\ (U\mathbb{T}, \Theta) &\mapsto U\Theta U^{-1} \end{aligned}$$

is a well-defined $n!$-to-1 mapping onto $\mathbb{U}(n)'$ with everywhere bijective differential $d\psi$, and if $\Theta = \mathbf{diag}(e^{i\theta_1}, \ldots, e^{i\theta_n})$, then

$$|\det d\psi_{(U\mathbb{T},\Theta)}| = \prod_{1 \le j < k \le n} |e^{i\theta_j} - e^{i\theta_k}|^2.$$

Proof To see that ψ is well defined, suppose that $U\mathbb{T} = \widetilde{U}\mathbb{T}$; i.e., that there is $\widetilde{\Theta} \in \mathbb{T}$ such that $U = \widetilde{U}\widetilde{\Theta}$. Then for any $\Theta \in \mathbb{T}'$,

$$U\Theta U^{-1} = \widetilde{U}\widetilde{\Theta}\Theta\widetilde{\Theta}^{-1}\widetilde{U}^{-1} = \widetilde{U}\Theta\widetilde{U}^{-1},$$

since the diagonal matrices Θ, $\widetilde{\Theta}$ commute.

Next, observe that if $U \in \mathbb{U}(n)'$ with distinct eigenvalues $\lambda_1, \ldots, \lambda_n$, then for any permutation $\sigma \in S_n$, there is $V_\sigma \in \mathbb{U}(n)$ such that

$$U = V_\sigma \, \mathbf{diag}(\lambda_{\sigma(1)}, \ldots, \lambda_{\sigma(n)}) V_\sigma^{-1},$$

and so U has (at least) $n!$ distinct preimages $(V_\sigma \mathbb{T}, \mathbf{diag}(\lambda_{\sigma(1)}, \ldots, \lambda_{\sigma(n)}))$. Moreover, if $V, W \in \mathbb{U}(n)$ are such that

$$V \, \mathbf{diag}(\lambda_{\sigma(1)}, \ldots, \lambda_{\sigma(n)}) V^{-1} = U = W \, \mathbf{diag}(\lambda_{\sigma(1)}, \ldots, \lambda_{\sigma(n)}) W^{-1},$$

then the k-th columns of both V and W are eigenvectors of U with eigenvalue $\lambda_{\sigma(k)}$. Since U has n distinct eigenvalues, its eigenspaces are one-dimensional, and so the k-th columns of V and W can only differ by multiplication by a unit modulus complex number ω_k. In other words,

$$W = V \, \mathbf{diag}(\omega_1, \ldots, \omega_n),$$

and since $\mathbf{diag}(\omega_1, \ldots, \omega_n) \in \mathbb{T}$, it follows that $V\mathbb{T} = W\mathbb{T}$. That is, U has exactly $n!$ preimages under ψ.

It remains to compute the differential $d\psi$.

A curve $\gamma : [0, 1] \to \mathbb{T}$ with $\gamma(0) = I$ has the form

$$\gamma(t) = \mathbf{diag}(e^{i\theta_1(t)}, \ldots, e^{i\theta_n(t)}),$$

where the $\theta_j(t)$ are arbitrary smooth functions of t, so that

$$\gamma'(0) = \mathbf{diag}(i\theta_1'(0), \ldots, i\theta_n'(0));$$

that is, $T_{I_n}(\mathbb{T}) \subseteq T_{I_n}(\mathbb{U}(n))$ is exactly the subspace of diagonal elements:

$$T_{I_n}(\mathbb{T}) = \{\mathbf{diag}(i\rho_1, \ldots, i\rho_n) : \rho_1, \ldots, \rho_n \in \mathbb{R}\} =: \mathfrak{t}.$$

Recall from Section 1.1 that the Lie algebra of $\mathbb{U}(n)$ itself is

$$\mathfrak{u}(n) = \left\{X \in \mathrm{M}_n(\mathbb{C}) : X + X^* = 0\right\}.$$

The map $\pi : \mathbb{U}(n) \to \mathbb{U}(n)/\mathbb{T}$ with $\pi(U) = U\mathbb{T}$ induces a surjective linear map $d\pi_{I_n} : T_{I_n}(\mathbb{U}(n)) \to T_{I_n\mathbb{T}}(\mathbb{U}(n)/\mathbb{T})$ whose kernel is exactly $\mathfrak{t}$, and so its image can be identified with the orthogonal complement of $\mathfrak{t}$ in $\mathfrak{u}$:

$$T_{I_n\mathbb{T}}(\mathbb{U}(n)/\mathbb{T}) \cong \{X \in \mathrm{M}_n(\mathbb{C}) : X + X^* = 0, X_{jj} = 0\,\forall j\} =: \mathfrak{p}.$$

Now, to compute $d\psi$ itself, we must compute $\frac{d}{dt}\psi \circ \gamma(t)|_{t=0}$ for smooth curves $\gamma : [0,1] \to (\mathbb{U}(n)/\mathbb{T}) \times \mathbb{T}'$ with arbitrary initial directions. We first consider the curve

$$\gamma_1(t) = (Ue^{tZ}\mathbb{T}, \Theta),$$

where Z is an $n \times n$ matrix over $\mathbb{C}$ with $Z + Z^* = 0$ and $Z_{jj} = 0$ for each j. Then γ_1 has $\gamma_1(0) = (U\mathbb{T}, \Theta)$ and $\gamma_1'(0) = (Z, 0)$ (it is customary to identify $T_{(U\mathbb{T},\Theta)}(\mathbb{U}(n)/\mathbb{T} \times \mathbb{T}')$ with $\mathfrak{p} \oplus \mathfrak{t}$),

$$\psi(\gamma_1(t)) = Ue^{tZ}\Theta e^{-tZ}U^{-1},$$

and

$$\begin{aligned} d\psi_{(U\mathbb{T},\Theta)}(Z,0) &= \frac{d}{dt}\psi(\gamma_1(t))|_{t=0} \\ &= UZ\Theta U^{-1} - U\Theta Z U^{-1} \\ &= \big[U\Theta U^{-1}\big]\big(U\big[\Theta^{-1}Z\Theta - Z\big]U^{-1}\big). \end{aligned}$$

In the orthogonal direction, consider the curve $\gamma_2 : [0,1] \to (\mathbb{U}(n)/\mathbb{T}) \times \mathbb{T}'$ defined by

$$\gamma_2(s) = (U\mathbb{T}, \Theta e^{s\Theta'}).$$

Then $\gamma_2(0) = (U\mathbb{T}, \Theta)$, $\gamma_2'(0) = (0, \Theta')$, and

$$d\psi_{(U\mathbb{T},\Theta)}(0,\Theta') = \frac{d}{ds}\psi \circ \gamma_2(s)|_{s=0} = U\Theta\Theta' U^{-1} = U\Theta U^{-1}\big(U\Theta' U^{-1}\big).$$

Together then,

$$d\psi_{(U\mathbb{T},\Theta)}(Z,\Theta') = U\Theta U^{-1}\big(U\big[\Theta^{-1}Z\Theta - Z + \Theta'\big]U^{-1}\big).$$

Identifying $T_{U\Theta U^{-1}}(\mathbb{U}(n)')$ with $\mathfrak{u}$, one would more typically write just

$$d\psi_{(U\mathbb{T},\Theta)}(Z,\Theta') = U\big[\Theta^{-1}Z\Theta - Z + \Theta'\big]U^{-1}.$$

Since multiplication by a unitary matrix is an isometry of $\mathrm{M}_n(\mathbb{C})$, it follows that

$$|\det \psi_{(U\mathbb{T},\Theta)}| = \left|\det A_\Theta \oplus I\right| = \left|\det A_\Theta\right|,$$

where $A_\Theta(Z) = \Theta^{-1}Z\Theta - Z$.

To finally actually compute the determinant, it is easier to consider the complexification of $\mathfrak{p}$: the space $\mathfrak{p}$ itself is a real vector space, but if $(X_1, \ldots, X_N)$ is a basis, then we may instead consider the complex vector space

$$\mathfrak{p}_\mathbb{C} := \{z_1X_1 + \cdots + z_NX_n : z_1, \ldots, z_n \in \mathbb{C}\}.$$

The matrices $(X_1, \ldots, X_N)$ are then a basis of $\mathfrak{p}_\mathbb{C}$, and any linear map on $\mathfrak{p}$ can be extended to a linear map on $\mathfrak{p}_\mathbb{C}$, whose determinant is the same as that of the original map.

It is an easy exercise that the complexification $\mathfrak{p}_\mathbb{C}$ is simply those matrices in $\mathrm{M}_n(\mathbb{C})$ with zeroes on the diagonal; this space has orthonormal basis E_{jk} ($j \neq k$), with a one in the j-k position and zeroes elsewhere. With respect to this basis, the operator A_Θ is diagonal:

$$A_\Theta(E_{jk}) = \left(e^{i(\theta_j - \theta_k)} - 1\right)E_{jk}.$$

It follows that

$$|\det \psi_{(U\mathbb{T},\Theta)}| = \prod_{j \neq k} \left|e^{i(\theta_j - \theta_k)} - 1\right| = \prod_{1 \leq j < k \leq n} \left|e^{i\theta_j} - e^{i\theta_k}\right|^2,$$

which completes the proof of the lemma. □

We now complete the proof of Weyl's integration formula.

Proof of Theorem 3.1 Since ψ is an $n!$-to-1 local diffeomorphism, using ψ to make a change of variables gives

$$\int_{\mathbb{U}(n)'} g(U) d\mu_{\mathbb{U}(n)'}(U)$$
$$= \frac{1}{n!} \int_{\mathbb{T}'} \int_{\mathbb{U}(n)/\mathbb{T}} g(\psi(U\mathbb{T}, \Theta)) |\det d\psi(U\mathbb{T}, \Theta)| d\mu_{\mathbb{U}(n)/\mathbb{T}}(U\mathbb{T}) d\mu_{\mathbb{T}'}(\Theta),$$

where each of the measures is the invariant measure on the appropriate space, as described in Proposition 3.2.

It is easy to see that $\mathbb{T} \setminus \mathbb{T}'$ and $\mathbb{U}(n) \setminus \mathbb{U}(n)'$ have measure zero, since their dimensions are strictly smaller than those of the full groups, and so we may instead write

$$\int_{\mathbb{U}(n)} g(U) d\mu_{\mathbb{U}(n)}(U)$$
$$= \frac{1}{n!} \int_{\mathbb{T}} \int_{\mathbb{U}(n)/\mathbb{T}} g(\psi(U\mathbb{T}, \Theta)) |\det d\psi(U\mathbb{T}, \Theta)| d\mu_{\mathbb{U}(n)/\mathbb{T}}(U\mathbb{T}) d\mu_{\mathbb{T}}(\Theta).$$

From Lemma 3.4,

$$|\det d\psi_{(U\mathbb{T},\Theta)}| = \prod_{1 \leq j < k \leq n} |e^{i\theta_j} - e^{i\theta_k}|^2,$$

and since g was assumed to be a class function,

$$g(\psi(U\mathbb{T}, \Theta)) = g(U\Theta U^{-1}) = g(\Theta) = \tilde{g}(\theta_1, \ldots, \theta_n),$$

where $\tilde{g}$ is as in the statement of the theorem and $\Theta = \mathbf{diag}(e^{i\theta_1}, \ldots, e^{i\theta_n})$.

Since $\mu_{\mathbb{U}(n)/\mathbb{T}}$ is a probability measure, and $d\mu_{\mathbb{T}} = \frac{1}{(2\pi)^n} d\theta_1 \cdots d\theta_n$, this completes the proof. □

The factor $\prod_{1 \leq j < k \leq n} |e^{i\theta_j} - e^{i\theta_k}|^2$ is the norm-squared of a Vandermonde determinant. Observe that for any given pair (j, k), $|e^{i\theta_k} - e^{i\theta_j}|^2$ is zero if $\theta_j = \theta_k$ (and small if they are close), but $|e^{i\theta_k} - e^{i\theta_j}|^2$ is 4 if $\theta_j = \theta_k + \pi$. This produces the effect alternatively known as *eigenvalue repulsion* or *eigenvalue rigidity*: each pair of eigenvalues repel each other, so that the collection of points is very evenly spaced. This is clearly visible in simulations, even for relatively small matrices. In the picture on the right in Figure 3.1, 80 points were dropped uniformly and independently (thus there is no repulsion); there are several large clumps of points close together, and some largeish gaps. The picture on the right is of the eigenvalues of a random 80×80 unitary matrix; one can clearly see that they are more regularly spaced around the circle.

There are integration formulae for the other matrix groups as well, although some details are slightly more complicated. First, recall the trivial eigenvalues: each matrix in $\mathbb{SO}(2n+1)$ has 1 as an eigenvalue, each matrix in $\mathbb{SO}^-(2n+1)$ has -1 as an eigenvalue, and each matrix in $\mathbb{SO}^-(2n+2)$ has both -1 and 1 as eigenvalues. The remaining eigenvalues of matrices in $\mathbb{SO}(n)$ or $\mathbb{S}\mathbb{p}(2n)$ occur in complex conjugate pairs. For this reason, when discussing $\mathbb{SO}(n)$, $\mathbb{SO}^-(n)$, or $\mathbb{S}\mathbb{p}(2n)$, the eigenvalue angles corresponding to the eigenvalues in the open upper half-circle are considered to be the nontrivial ones, and one normally considers the eigenvalue process restricted to that set. For $\mathbb{U}(n)$, all the eigenvalue angles are considered nontrivial; there are no automatic symmetries in this case.

The following result gives the analogues of the formula in Theorem 3.1 for the remaining groups.

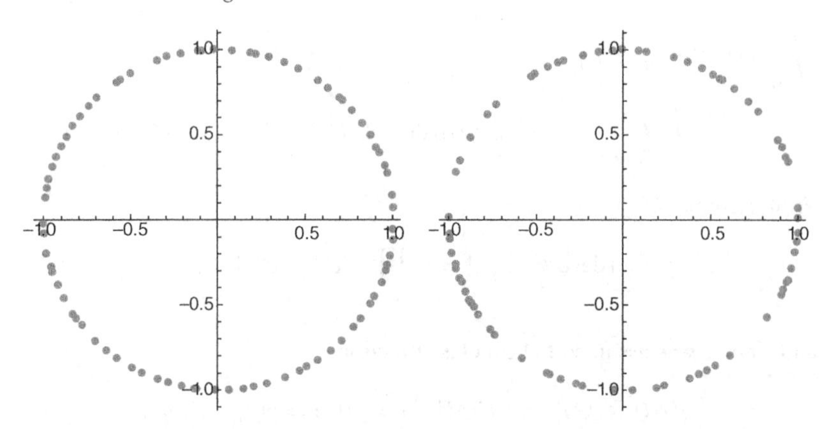

Figure 3.1 On the left are the eigenvalues of an 80×80 random unitary matrix; on the right are 80 i.i.d. uniform random points.

Theorem 3.5 *Let U be a Haar-distributed random matrix in S, where S is one of $\mathbb{SO}(2n+1)$, $\mathbb{SO}(2n)$, $\mathbb{SO}^-(2n+1)$, $\mathbb{SO}^-(2n+2)$, $\mathbb{Sp}(2n)$. Then a function g of U that is invariant under conjugation of U by a fixed orthogonal (in all but the last case) or symplectic (in the last case) matrix is associated as above with a function $\tilde{g} : [0,\pi)^n \to \mathbb{R}$ (of the nontrivial eigenangles) that is invariant under permutations of coordinates, and if U is distributed according to Haar measure on G, then*

$$\mathbb{E}g(U) = \int_{[0,\pi)^n} \tilde{g}\, d\mu_G^W,$$

where the measures μ_G^W on $[0,\pi)^n$ have densities with respect to $d\theta_1 \cdots d\theta_n$ as follows.

G	μ_G^W
$\mathbb{SO}(2n)$	$\dfrac{2}{n!\,(2\pi)^n} \prod_{1\le j<k\le n} \left(2\cos(\theta_k) - 2\cos(\theta_j)\right)^2$
$\mathbb{SO}(2n+1)$	$\dfrac{2^n}{n!\,\pi^n} \prod_{1\le j\le n} \sin^2\left(\dfrac{\theta_j}{2}\right) \prod_{1\le j<k\le n} \left(2\cos(\theta_k) - 2\cos(\theta_j)\right)^2$
$\mathbb{SO}^-(2n+1)$	$\dfrac{2^n}{n!\,\pi^n} \prod_{1\le j\le n} \cos^2\left(\dfrac{\theta_j}{2}\right) \prod_{1\le j<k\le n} \left(2\cos(\theta_k) - 2\cos(\theta_j)\right)^2$
$\mathbb{Sp}(2n)$ $\mathbb{SO}^-(2n+2)$	$\dfrac{2^n}{n!\,\pi^n} \prod_{1\le j\le n} \sin^2(\theta_j) \prod_{1\le j<k\le n} \left(2\cos(\theta_k) - 2\cos(\theta_j)\right)^2$

So as not to spoil the reader's fun, we will show how to modify the proof of the unitary case for the case of $\mathbb{SO}(2n)$ and leave the remaining formulae as exercises.

Proof of the Weyl integration formula for $\mathbb{SO}(2n)$ If $U \in \mathbb{SO}(2n)$, then there are $V \in \mathbb{SO}(2n)$ and angles $\{\theta_1, \ldots, \theta_n\} \subseteq [0, \pi]$ such that

$$U = V \begin{bmatrix} \cos\theta_1 & \pm\sin(\theta_1) & & & \\ \mp\sin(\theta_1) & \cos(\theta_1) & & & \\ & & \ddots & & \\ & & & \cos(\theta_n) & \pm\sin(\theta_n) \\ & & & \mp\sin(\theta_n) & \cos(\theta_n) \end{bmatrix} V^{-1}; \quad (3.2)$$

the eigenvalues of U are the complex conjugate pairs $\{e^{\pm i\theta_1}, \ldots, e^{\pm i\theta_n}\}$.

Let $\mathbb{T}$ denote the set of all block diagonal rotations as in (3.2), $\mathbb{T}' \subset \mathbb{T}$ those matrices for which the θ_j are all distinct and different from 0 and π, and let $\mathbb{SO}(2n)' \subseteq \mathbb{SO}(2n)$ be the subset of $\mathbb{SO}(2n)$ with distinct eigenvalues, different from 1 and -1. Define a map

$$\begin{aligned} \psi : (\mathbb{SO}(2n)/\mathbb{T}) \times \mathbb{T}' &\to \mathbb{SO}(2n)' \\ (U\mathbb{T}, R) &\mapsto URU^{-1}. \end{aligned}$$

Then ψ is a $2^{n-1}n!$-to-1 covering of $\mathbb{SO}(2n)'$: each $U \in \mathbb{SO}(2n)$ has $2^{n-1}n!$ distinct preimages, corresponding to the $n!$ possible permutations of the 2×2 blocks and the 2^{n-1} possible reorderings of the bases within all but the last of the corresponding 2-dimensional subspaces (the last one is then forced so as to remain within $\mathbb{SO}(2n)$).

As before, ψ is used to make a change of variables, and the crucial ingredient in the integration formula is the determinant of $d\psi$. Recall that

$$\mathfrak{so}(2n) := T_{I_n}(\mathbb{SO}(2n)) = \{X \in \mathrm{M}_{2n}(\mathbb{R}) : X + X^T = 0\}.$$

If $\gamma : [0, 1] \to \mathbb{T}$ has the form

$$\gamma(s) = \begin{bmatrix} \cos(\theta_1(s)) & \pm\sin(\theta_1(s)) & & & \\ \mp\sin(\theta_1(s)) & \cos(\theta_1(s)) & & & \\ & & \ddots & & \\ & & & \cos(\theta_n(s)) & \pm\sin(\theta_n(s)) \\ & & & \mp\sin(\theta_n(s)) & \cos(\theta_n(s)) \end{bmatrix}$$

(with the choice of signs consistent as s varies), then

$$\gamma'(0) = \begin{bmatrix} 0 & \pm\theta_1'(0) & & & \\ \mp\theta_1'(0) & 0 & & & \\ & & \ddots & & \\ & & & 0 & \pm\theta_n'(0) \\ & & & \mp\theta_n'(0) & 0 \end{bmatrix},$$

and so

$$\mathfrak{t} := T_{I_n}(\mathbb{T}) = \left\{ \begin{bmatrix} 0 & \rho_1 \\ -\rho_1 & 0 \end{bmatrix} \oplus \cdots \oplus \begin{bmatrix} 0 & \rho_n \\ -\rho_n & 0 \end{bmatrix} : \rho_1, \ldots, \rho_n \in \mathbb{R} \right\}.$$

The map $\pi : \mathbb{SO}(2n) \to \mathbb{SO}(2n)/\mathbb{T}$ with $\pi(U) := U\mathbb{T}$ induces a surjective linear map $d\pi_{I_n} : T_{I_n}(\mathbb{SO}(2n)) \to T_{I_n\mathbb{T}}(\mathbb{SO}(2n)/\mathbb{T})$ with kernel $\mathfrak{t}$, whose image is therefore isomorphic to the orthogonal complement

$$\mathfrak{p} := \left\{ X \in \mathrm{M}_{2n}(\mathbb{R}) : X + X^T = 0, \begin{bmatrix} X_{2j-1,2j-1} & X_{2j-1,2j} \\ X_{2j,2j-1} & X_{2j,2j} \end{bmatrix} = \begin{bmatrix} 0 & 0 \\ 0 & 0 \end{bmatrix}, j = 1, \ldots, n \right\}.$$

Computing the tangent vectors at 0 to curves $\gamma_1(t) = (Ue^{tZ}\mathbb{T}, R)$ and $\gamma_2(s) = (U\mathbb{T}, Re^{sR'})$ as in the unitary case shows that

$$d\psi_{(U\mathbb{T},R)}(Z, R') = U\big[(A_R \oplus I)(Z, R)\big]U^{-1},$$

where $A_R : \mathfrak{p} \to \mathfrak{p}$ is given by $A_R(Z) = R^{-1}ZR - Z$, and we have identified $T_{(U\mathbb{R},R)}(\mathbb{SO}(n)/\mathbb{T} \times \mathbb{T}')$ with $\mathfrak{p} \oplus \mathfrak{t}$ and $T_{URU^{-1}}(\mathbb{SO}(n)')$ with $\mathfrak{so}(2n)$. We thus have that

$$|\det d\psi_{(U\mathbb{T},R)}| = |\det A_R|.$$

Unfortunately, there is no obvious basis in which A_R is diagonal, and so computing $|\det A_R|$ requires a few linear-algebraic tricks. First, it is once again simpler to compute in the complexification

$$\mathfrak{p}_{\mathbb{C}} = \left\{ X \in \mathrm{M}_{2n}(\mathbb{C}) : X + X^T = 0, \begin{bmatrix} X_{2j-1,2j-1} & X_{2j-1,2j} \\ X_{2j,2j-1} & X_{2j,2j} \end{bmatrix} = \begin{bmatrix} 0 & 0 \\ 0 & 0 \end{bmatrix}, j = 1, \ldots, n \right\},$$

of $\mathfrak{p}$. Note that while in the unitary case, complexifying $\mathfrak{p}$ removed the condition $X + X^* = 0$, in the orthogonal case, the condition $X + X^T = 0$ remains.

Next, we change bases on $\mathbb{C}^{2n}$ to diagonalize the elements of $\mathbb{T}$: let

$$C := \begin{bmatrix} 1 & 1 \\ i & -i \end{bmatrix} \oplus \cdots \oplus \begin{bmatrix} 1 & 1 \\ i & -i \end{bmatrix},$$

and let C act on $\mathrm{M}_n(\mathbb{C})$ by conjugation. The subspace $\mathfrak{t}$, and hence also $\mathfrak{p}_{\mathbb{C}}$, is preserved by this action, and

$$C^{-1}\left(\begin{bmatrix} \cos(\theta_1) & \sin(\theta_1) \\ -\sin(\theta_1) & \cos(\theta_1) \end{bmatrix} \oplus \cdots \oplus \begin{bmatrix} \cos(\theta_n) & \sin(\theta_n) \\ -\sin(\theta_n) & \cos(\theta_n) \end{bmatrix}\right)C$$
$$= \begin{bmatrix} e^{i\theta_1} & 0 \\ 0 & e^{-i\theta_1} \end{bmatrix} \oplus \cdots \oplus \begin{bmatrix} e^{i\theta_n} & 0 \\ 0 & e^{-i\theta_n} \end{bmatrix}.$$

Now, $\mathfrak{p}_{\mathbb{C}}$ has the obvious basis F_{jk}, where F_{jk} has a 1 in the j-k entry, a -1 in the k-j entry and zeroes otherwise (where (j,k), runs over pairs with $j < k$ which are not in any of the 2×2 blocks along the diagonal). However, A_R is not diagonal with respect to the basis $CF_{jk}C^{-1}$, and one more trick is in order.

Observe that if $S : \mathrm{M}_n(\mathbb{C}) \to \mathrm{M}_n(\mathbb{C})$ is defined by

$$[SX]_{jk} = \begin{cases} X_{jk}, & j \le k; \\ -X_{jk}, & j > k, \end{cases}$$

then S commutes with A_R (this is easy to check on the basis $CE_{jk}C^{-1}$ of $\mathrm{M}_n(\mathbb{C})$).

Exercise 3.6 Let V be a finite-dimensional inner product space and $U \subseteq V$ a subspace. Let $T : V \to V$ such that $T(U) = U$ and let $S : V \to V$ commute with T. Then $T(S(U)) = S(U)$ and

$$\det T|_U = \det T|_{S(U)}.$$

Applying the exercise to the maps $T = A_R$ and S defined above, it follows that our quarry $\det A_R$, when A_R is viewed as a map on $\mathfrak{p}_{\mathbb{C}}$, is the same as $\det A_R$ when A_R is viewed as a map on

$$\mathfrak{q}_{\mathbb{C}} := \left\{X \in \mathrm{M}_{2n}(\mathbb{C}) : X - X^T = 0, \begin{bmatrix} X_{2j-1,2j-1} & X_{2j-1,2j} \\ X_{2j,2j-1} & X_{2j,2j} \end{bmatrix} = \begin{bmatrix} 0 & 0 \\ 0 & 0 \end{bmatrix}, j = 1, \ldots, n\right\}.$$

The subspace $\mathfrak{q}_{\mathbb{C}}$ has basis G_{jk} defined similarly to F_{jk} but with both nonzero entries equal to 1, and so in particular the subspaces $\mathfrak{p}_{\mathbb{C}}$ and $\mathfrak{q}_{\mathbb{C}}$ are orthogonal. It follows that $\det A_R$ when A_R is viewed as a map on $\mathfrak{p}_{\mathbb{C}} \oplus \mathfrak{q}_{\mathbb{C}}$ is the square of

$\det A_R$ when A_R is restricted to $\mathfrak{p}_{\mathbb{C}}$. The point, of course, is that $\mathfrak{p}_{\mathbb{C}} \oplus \mathfrak{q}_{\mathbb{C}}$ has basis E_{jk}, where $j \neq k$ and (j,k) is not in any of the 2×2 blocks along the diagonal, and with respect to $CE_{jk}C^{-1}$, A_R is diagonal: if $D = C^{-1}RC$ is the diagonalization of R by C, then

$$\begin{aligned} A_R(CE_{jk}C^{-1}) &= CD^{-1}C^{-1}(CE_{jk}C^{-1})CDC^{-1} - CE_{jk}C^{-1} \\ &= \left(\exp\left\{i\left((-1)^j\theta_{\lceil\frac{j}{2}\rceil} - (-1)^k\theta_{\lceil\frac{k}{2}\rceil}\right)\right\} - 1\right) CE_{jk}C^{-1}. \end{aligned}$$

Examining the coefficient above, one can see that for each pair $p < q$, each of the four factors $e^{i(\pm\theta_p \pm\theta_q)} - 1$ appears as an eigenvalue exactly twice. Since

$$|e^{i(\theta_p+\theta_q)} - 1||e^{i(\theta_p-\theta_q)} - 1| = |2\cos(\theta_p) - 2\cos(\theta_q)|,$$

we finally have that

$$|\det_{\mathfrak{p}_{\mathbb{C}}\oplus\mathfrak{q}_{\mathbb{C}}} d\psi_{(U\mathbb{T},R)}| = \prod_{1\le p<q\le n} |2\cos(\theta_p) - 2\cos(\theta_q)|^4,$$

and so

$$|\det_{\mathfrak{p}_{\mathbb{C}}} d\psi_{(U\mathbb{T},R)}| = \prod_{1\le p<q\le n} |2\cos(\theta_p) - 2\cos(\theta_q)|^2. \tag{3.3}$$

To complete the proof, let $g : \mathbb{SO}(2n) \to \mathbb{R}$ be a class function; since ψ is a $\frac{2^n n!}{2}$-to-1 local diffeomorphism,

$$\begin{aligned} &\int_{\mathbb{SO}(2n)'} g(U) d\mu_{\mathbb{SO}(2n)'}(U) \\ &= \frac{2}{2^n n!}\int_{\mathbb{T}'}\int_{\mathbb{SO}(2n)/\mathbb{T}} g(\psi(U\mathbb{T},R))|\det d\psi_{(U\mathbb{T},R)}| d\mu_{\mathbb{SO}(n)/\mathbb{T}}(U\mathbb{T}) d\mu_{\mathbb{T}'}(R), \end{aligned}$$

where all of the measures are the (normalized) Haar measures on the appropriate spaces. As before, the excluded parts of $\mathbb{SO}(2n)$ and $\mathbb{T}$ are lower-dimensional and hence of measure zero, and so we have

$$\begin{aligned} &\int_{\mathbb{SO}(2n)} g(U) d\mu_{\mathbb{SO}(2n)}(U) \\ &= \frac{2}{2^n n!}\int_{\mathbb{T}}\int_{\mathbb{SO}(2n)/\mathbb{T}} g(\psi(U\mathbb{T},R))|\det d\psi_{(U\mathbb{T},R)}| d\mu_{\mathbb{SO}(n)/\mathbb{T}}(U\mathbb{T}) d\mu_{\mathbb{T}}(R). \end{aligned}$$

Since g is a class function,

$$g(\psi(U\mathbb{T}, R)) = g(URU^{-1}) = g(R);$$

together with the expression for $|\det d\psi_{(U\mathbb{T},R)}|$ in (3.3), this completes the proof. □

3.2 Determinantal Point Processes

One important consequence of the Weyl formulae is that the eigenvalue processes of the classical compact groups are what are known as *determinantal point processes*, defined as follows.

Definition A **point process** $\mathcal{X}$ in a locally compact Polish space Λ is a random discrete subset of Λ. For $A \subseteq \Lambda$, $\mathcal{N}_A$ denotes the (random) number of points of $\mathcal{X}$ in A. The function $A \mapsto \mathcal{N}_A$ is called the **counting function** of $\mathcal{X}$.

Let μ be a Borel measure on Λ. For a point process $\mathcal{X}$ in Λ, if there exist functions $\rho_k : \Lambda^k \to [0, \infty)$ such that, for pairwise disjoint subsets $A_1, \ldots, A_k \subseteq \Lambda$,

$$\mathbb{E}\left[\prod_{j=1}^{k} \mathcal{N}_{A_j}\right] = \int_{A_1} \cdots \int_{A_k} \rho_k(x_1, \ldots, x_k) d\mu(x_1) \cdots d\mu(x_k),$$

then the ρ_k are called the k**-point correlation functions** (or **joint intensities**) of $\mathcal{X}$, with respect to μ.

A determinantal point process is a point process whose k-point correlation functions have a special form:

Definition Let $K : \Lambda \times \Lambda \to \mathbb{C}$. A point process $\mathcal{X}$ is a **determinantal point process with respect to** μ **with kernel** K if for all $k \in \mathbb{N}$,

$$\rho_k(x_1, \ldots, x_k) = \det\left[K(x_i, x_j)\right]_{i,j=1}^{k}.$$

It is a result of Macchi [73] and Soshnikov [94] that if a kernel K defines an operator $\mathcal{K}$ by

$$\mathcal{K}f(x) = \int_{\Lambda} K(x, y) f(y) d\mu(y)$$

and that operator is self-adjoint and trace class, then there is a determinantal point process on Λ with kernel K if and only if $\mathcal{K}$ has all eigenvalues in $[0, 1]$.

Proposition 3.7 *The nontrivial eigenvalue angles of uniformly distributed random matrices in any of* $\mathbb{SO}(N)$, $\mathbb{SO}^{-}(N)$, $\mathbb{U}(N)$, $\mathbb{Sp}(2N)$ *are a determinantal* *point process with respect to uniform (probability) measure on Λ, with kernels as follows.*

	$K_N(x,y)$	Λ
$\mathbb{U}(N)$	$\sum_{j=0}^{N-1} e^{ij(x-y)}$	$[0,2\pi)$
$\mathbb{SO}(2N)$	$1+\sum_{j=1}^{N-1} 2\cos(jx)\cos(jy)$	$[0,\pi)$
$\mathbb{SO}(2N+1)$	$\sum_{j=0}^{N-1} 2\sin\left(\frac{(2j+1)x}{2}\right)\sin\left(\frac{(2j+1)y}{2}\right)$	$[0,\pi)$
$\mathbb{SO}^-(2N+1)$	$\sum_{j=0}^{N-1} 2\cos\left(\frac{(2j+1)x}{2}\right)\cos\left(\frac{(2j+1)y}{2}\right)$	$[0,\pi)$
$\mathbb{S}\mathrm{p}(2N)$ $\mathbb{SO}^-(2N+2)$	$\sum_{j=1}^{N} 2\sin(jx)\sin(jy)$	$[0,\pi)$

In order to prove the proposition, we first describe a common framework in which each of the groups can be treated.

Equip Λ with a finite measure μ, and suppose that $f:\Lambda\to\mathbb{C}$ is such that

$$\int f d\mu = 0 \qquad \int |f|^2 d\mu = 1.$$

Suppose that $p_n(x)$ are monic polynomials such that the sequence $\varphi_0 := \frac{1}{\sqrt{\mu(\Lambda)}}$, $\varphi_n = p_n(f)$ $(n \geq 1)$ is orthonormal in $L_2(\mu)$; i.e.,

$$\int_\Lambda \varphi_j \overline{\varphi}_k d\mu = \delta_{jk}.$$

Define the function K_N on Λ^2 by

$$K_N(x,y) := \sum_{j=0}^{N-1} \varphi_j(x)\overline{\varphi}_j(y),$$

and for each $m \geq 1$, define $D_{m,N} : \Lambda^m \to \mathbb{C}$ by

$$D_{m,N}(x_1,\ldots,x_m) := \det\left[K_N(x_j,x_k)\right]_{j,k=1}^m.$$

Take $D_{0,N} = 1$.

Lemma 3.8 *1. For all* $(x_1, \ldots, x_N) \in \Lambda^N$,

$$D_{N,N} = \frac{1}{\mu(\Lambda)} \left| \det \left[f(x_j)^{k-1} \right]_{j,k=1}^N \right|^2.$$

In particular, $D_{N,N}$ *is real-valued, nonnegative, and symmetric in* $(x_1, \ldots, x_N)$.

2. *For* $1 \leq n \leq N$,

$$\int_\Lambda D_{n,N}(x_1, \ldots, x_n) d\mu(x_n) = (N + 1 - n) D_{n-1,N}(x_1, \ldots, x_{n-1}).$$

The function $D_{n,N}$ *is real-valued, nonnegative, and symmetric in its arguments.*

3. *For* $m > N$, $D_{m,N} = 0$.
4. *Let* $g : \Lambda^n \to \mathbb{R}$. *Then*

$$\frac{1}{n!} \int_{\Lambda^n} g(x_1, \ldots, x_n) D_{n,N}(x_1, \ldots, x_n) d\mu(x_1) \cdots d\mu(x_n)$$
$$= \frac{1}{N!} \int_{\Lambda^N} \left(\sum_{1 \leq j_1 < \cdots < j_n \leq N} g(x_{j_1}, \ldots, x_{j_n}) \right) D_{N,N}(x_1, \ldots, x_N) d\mu(x_1) \cdots d\mu(x_N).$$

The value of the lemma for us is that it is not too hard to check, using the Weyl integration formulae, that the N-point correlation functions are those claimed in Proposition 3.7; the full determinantal structure then follows from the lemma.

Proof

1. Observe that since $\varphi_n = p_n(f)$ for a monic polynomial p_n, starting from the last column and adding linear combinations of previous columns, then multiplying the first column by $\frac{1}{\sqrt{\mu(\Lambda)}}$,

$$\frac{1}{\sqrt{\mu(\Lambda)}} \det \left[f(x_j)^{k-1} \right]_{j,k=1}^N = \det \left[\varphi_{k-1}(x_j) \right]_{j,k=1}^N,$$

and so

$$\begin{aligned}
\frac{1}{\mu(\Lambda)} \left| \det \left[f(x_j)^{k-1} \right]_{j,k=1}^N \right|^2 &= \left(\det \left[\varphi_{k-1}(x_j) \right]_{j,k=1}^N \right) \left(\det \left[\overline{\varphi}_{k-1}(x_j) \right]_{j,k=1}^N \right) \\
&= \left(\det \left[\varphi_{k-1}(x_j) \right]_{j,k=1}^N \right) \left(\det \left[\overline{\varphi}_{j-1}(x_k) \right]_{j,k=1}^N \right) \\
&= \det \left[\sum_{\ell=1}^N \varphi_{\ell-1}(x_j) \overline{\varphi}_{\ell-1}(x_k) \right]_{j,k=1}^N \\
&= D_{N,N}(x_1, \ldots, x_N).
\end{aligned}$$

2. We prove the statement by induction, using the Laplace expansion of the determinant. The $n = 1$ case is trivial by the orthonormality of the φ_j.

   For $n > 1$, let $A_{n,N} := \left[K_N(x_j, x_k)\right]_{j,k=1}^n$ and expand $\det A_{n,N}$ along the final column:

$$\det A_{n,N} = \sum_{k=1}^{n} (-1)^{k+n} K_N(x_k, x_n) \det A_{n,N}^{k,n},$$

where $A_{n,N}^{k,\ell}$ denotes the matrix $A_{n,N}$ with the kth row and ℓ^{th} column removed.

   When $k = n$, the matrix $A_{n,N}^{k,n}$ is just $\left[K_N(x_j, x_k)\right]_{j,k=1}^{n-1}$, which is in particular independent of x_n, and integrating $K_N(x_n, x_n)$ produces a factor of N.

   For the remaining terms, expanding $\det A_{n,N}^{k,n}$ along the bottom row gives

$$\det A_{n,N}^{k,n} = \sum_{\ell=1}^{n-1} (-1)^{\ell+n-1} K_N(x_n, x_\ell) \det \left[K_N(x_i, x_j)\right]_{\substack{1 \le i,j \le n \\ i \neq k,n; j \neq \ell,n}}.$$

Note in particular that there are no x_n's left in the determinant.

   Now,

$$\begin{aligned} \int_\Lambda K_N(x_k, x_n) K_N(x_n, x_\ell) d\mu(x_n) &= \sum_{r,s=1}^{N-1} \varphi_r(x_k) \overline{\varphi}_s(x_\ell) \int_\Lambda \overline{\varphi}_r(x_n) \varphi_s(x_n) d\mu(x_n) \\ &= \sum_{r=0}^{N-1} \varphi_r(x_k) \overline{\varphi}_r(x_\ell) \\ &= K_N(x_k, x_\ell). \end{aligned}$$

   It follows that for $k < n$,

$$\begin{aligned} &\int_\Lambda (-1)^{k+n} K_N(x_k, x_n) \det A_{n,N}^{k,n} d\mu(x_n) \\ &\quad = \sum_{\ell=1}^{n-1} (-1)^{k+\ell-1} K_N(x_k, x_\ell) \det \left[K_N(x_i, x_j)\right]_{\substack{1 \le i,j \le n \\ i \neq k,n; j \neq \ell,n}} \\ &\quad = -\det \left[K_N(x_i, x_j)\right]_{i,j=1}^{n-1}. \end{aligned}$$

As there are $n - 1$ terms of this type, together with the $k = n$ case this gives the formula in part 2.

   Since we have already seen that $D_{N,N}$ is real-valued, nonnegative, and symmetric, it follows immediately from the formula just established that $D_{n,N}$ is as well.

3. Using the same trick as in part 1., if $n > N$, then
$$\det\left[K_N(x_i, x_j)\right]_{i,j=1}^n = \det\left[\Phi\Phi^*\right],$$
where Φ is an $n \times N$ matrix with entries $\varphi_{j-1}(x_i)$. Since the rank of $\Phi\Phi^*$ is at most $N < n$, $\det \Phi\Phi^* = 0$.
4. The only nontrivial case is $n < N$. By the symmetry of $D_{N,N}(x_1, \ldots, x_N)$,
$$\int_{\Lambda^N} \left(\sum_{1 \le j_1 \cdots < j_n \le N} g(x_{j_1}, \ldots, x_{j_n}) \right) D_{N,N}(x_1, \ldots, x_N) d\mu(x_N) \cdots d\mu(x_1)$$
$$= \binom{N}{n} \int_{\Lambda^N} g(x_1, \ldots, x_n) D_{N,N}(x_1, \ldots, x_N) d\mu(x_N) \cdots d\mu(x_1).$$
Using part 2 to integrate out $x_N, \ldots, x_{n+1}$ gives that this is
$$\frac{N!}{n!} \int_{\Lambda^n} g(x_1, \ldots, x_n) D_{n,N}(x_1, \ldots, x_n) d\mu(x_n) \cdots d\mu(x_1).$$
□

Proof of Proposition 3.7 *Unitary case:* Let $\mathcal{X} = \left\{e^{i\theta_1}, \ldots, e^{i\theta_N}\right\}$ be the eigenvalue process corresponding to Haar measure on the unitary group. Let $\Lambda = [0, 2\pi)$, let $f : \Lambda \to \mathbb{C}$ be defined by $f(\theta) = e^{i\theta}$, and for $n \ge 1$, let $\varphi_n = f^n$. Then the kernel $K_N(x, y)$ from the lemma is
$$K_N(x, y) = \sum_{j=0}^{N-1} \varphi_j(x)\overline{\varphi}_j(y) = \sum_{j=0}^{N-1} e^{ij(x-y)}$$
and the Weyl density is
$$\prod_{1 \le j < k \le N} |e^{i\theta_j} - e^{i\theta_k}|^2 = \left| \det\left[e^{i\theta_j(k-1)}\right]_{j,k=1}^N \right|^2 = D_{N,N},$$
by part 1 of the lemma.

Now, if $A_1, \ldots, A_N \subseteq \Lambda$ are pairwise disjoint, then $\prod_{j=1}^N \mathcal{N}_{A_j}(\mathcal{X}) = 1$ if there is some permutation $\sigma \in S_N$ such that $\theta_j \in A_{\sigma(j)}$ for each j, and otherwise $\prod_{j=1}^N \mathcal{N}_{A_j}(\mathcal{X}) = 0$. It thus follows from the Weyl integration formula that
$$\mathbb{E}\left[\prod_{j=1}^N \mathcal{N}_{A_j}\right] = N!\,\mathbb{P}\left[e^{i\theta_1} \in A_1, \ldots, e^{i\theta_N} \in A_N\right]$$
$$= \frac{1}{(2\pi)^N} \int_{A_1} \cdots \int_{A_N} \prod_{1 \le j < k \le N} |e^{i\theta_j} - e^{i\theta_k}|^2 d\theta_N \cdots d\theta_1$$
$$= \frac{1}{(2\pi)^N} \int_{A_1} \cdots \int_{A_N} D_{N,N}(\theta_1, \ldots, \theta_N) d\theta_N \cdots d\theta_1.$$

More generally, if $A_1, \ldots, A_k \subseteq \Lambda$ are pairwise disjoint, then

$$\mathbb{E}\left[\prod_{j=1}^{k} \mathcal{N}_{A_j}\right] = \mathbb{E}\left[\prod_{j=1}^{k}\left(\sum_{\ell=1}^{N} \mathbb{1}_{A_j}(\theta_\ell)\right)\right]$$
$$= \sum_{\sigma \in S_k} \sum_{1 \le \ell_1 < \cdots < \ell_k \le N} \mathbb{E}\left[\prod_{j=1}^{k} \mathbb{1}_{A_{\sigma(j)}}(\theta_{\ell_j})\right]$$
$$= \frac{k!}{N!\,(2\pi)^N} \int_{\Lambda^N} \left(\sum_{1 \le \ell_1 < \cdots < \ell_k \le N} \left[\prod_{j=1}^{k} \mathbb{1}_{A_j}(\theta_{\ell_j})\right]\right)$$
$$\times D_{N,N}(\theta_1, \ldots, \theta_N) d\theta_1 \cdots d\theta_N.$$

By part 4 of the Lemma, this is

$$\frac{1}{(2\pi)^k} \int_{\Lambda^k} \left(\prod_{j=1}^{k} \mathbb{1}_{A_j}(\theta_j)\right) D_{k,N}(\theta_1, \ldots, \theta_k) d\theta_1 \cdots d\theta_k$$
$$= \frac{1}{(2\pi)^k} \int_{A_1} \cdots \int_{A_k} \det\left[K_N(\theta_i, \theta_j)\right]_{i,j=1}^{k} d\theta_N \cdots d\theta_1.$$

Even special orthogonal case: Let $\mathcal{X} = \left\{e^{i\theta_1}, \ldots, e^{i\theta_N}\right\}$ be the nontrivial eigenvalue process corresponding to Haar measure on $\mathbb{SO}(2N)$, and let $\Lambda = [0, \pi]$. In order to work within the framework of the lemma, we choose μ such that $d\mu = \frac{1}{2\pi} d\theta$ on Λ, so that $\mu(\Lambda) = \frac{1}{2}$. Let $f : \Lambda \to \mathbb{C}$ be defined by $f(\theta) = 2\cos(\theta) = e^{i\theta} + e^{-i\theta}$; note that indeed

$$\int f d\mu = 0 \qquad \int |f|^2 d\mu = 1.$$

It is easy to show by induction that the function

$$\varphi_n(\theta) = 2\cos(n\theta) = e^{in\theta} + e^{-in\theta}$$

is a monic polynomial of degree n in $e^{i\theta} + e^{-i\theta}$, and if we choose $\varphi_0(\theta) = \sqrt{2}$, then $\{\varphi_n\}_{n \ge 0}$ is an orthonormal sequence with respect to μ. Indeed, the reason for the choice of normalization of μ is so that this sequence of monic polynomials in $2\cos(\theta)$ is orthonormal.

The kernel from the lemma corresponding to our choice of μ is

$$K_N^{\mu}(x, y) = \sum_{j=0}^{N-1} \varphi_j(x)\overline{\varphi}_j(y) = 2 + \sum_{j=1}^{N-1} 4\cos(jx)\cos(jy).$$

Now, the density given by the Weyl integration formula in the case of $\mathbb{SO}(2N)$ is

$$\frac{2}{N!\,(2\pi)^N}\prod_{1\le j<k\le N}(2\cos(\theta_j)-2\cos(\theta_k))^2$$
$$=\frac{1}{N!\,(2\pi)^N\mu(\Lambda)}\left|\det\left[\left(2\cos(\theta_j)\right)^{k-1}\right]_{j,k=1}^N\right|^2=\frac{1}{N!\,(2\pi)^N}D_{N,N},$$

by part 1 of the lemma. So if $A_1,\dots,A_k\subseteq\Lambda$ are pairwise disjoint, then as above,

$$\mathbb{E}\left[\prod_{j=1}^k\mathcal{N}_{A_j}\right]=\frac{k!}{N!\,(2\pi)^N}\int_{\Lambda^N}\left(\sum_{1\le\ell_1<\cdots<\ell_k\le N}\left[\prod_{j=1}^k\mathbb{1}_{A_j}(\theta_{\ell_j})\right]\right)$$
$$\times D_{N,N}(\theta_1,\dots,\theta_N)d\theta_1\cdots d\theta_N$$
$$=\frac{1}{(2\pi)^k}\int_{\Lambda^k}\left(\prod_{j=1}^k\mathbb{1}_{A_j}(\theta_j)\right)D_{k,N}d\theta_1\cdots d\theta_k$$
$$=\frac{1}{(2\pi)^k}\int_{A_1}\cdots\int_{A_k}\det\left[2+4\sum_{j=1}^{N-1}\cos(\theta_r)\cos(\theta_s)\right]_{r,s=1}^k d\theta_1\cdots d\theta_k.$$

Cancelling the $\frac{1}{2^k}$ in front of the integral with a factor of 2 in the matrix inside the determinant shows that the $\mathbb{SO}(2N)$ eigenvalue process is determinantal with respect to the uniform probability measure on $[0,\pi]$ with kernel

$$K_N(x,y)=1+2\sum_{j=1}^{N-1}\cos(jx)\cos(jy).$$

Odd special orthogonal case: Let $\mathcal{X}=\left\{e^{i\theta_1},\dots,e^{i\theta_N}\right\}$ be the nontrivial eigenvalue process corresponding to Haar measure on $\mathbb{SO}(2N+1)$, and let $\Lambda=[0,\pi]$. Note that the form of the Weyl density in this case is somewhat different:

$$\frac{2^N}{N!\,\pi^N}\prod_{1\le j\le N}\sin^2\left(\frac{\theta_j}{2}\right)\prod_{1\le j<k\le N}(2\cos(\theta_k)-2\cos(\theta_j))^2$$

contains a product over j in addition to the Vandermonde factor. Rather than treating this product as a determinant, it is more convenient within the framework of Lemma 3.8 to treat the $\sin^2\left(\frac{\theta_j}{2}\right)$ factors as defining the reference measure μ. That is, let μ be the probability measure with $d\mu=\frac{2}{\pi}\sin^2\left(\frac{\theta}{2}\right)d\theta$ on $[0,\pi]$.

The Vandermonde factor in the Weyl density suggests taking $f(\theta) = 2\cos(\theta)$; since

$$\frac{2}{\pi}\int_0^{\pi} 2\cos(\theta)\sin^2\left(\frac{\theta}{2}\right)d\theta = -1,$$

we take instead

$$f(\theta) = 1 + 2\cos(\theta) = \frac{\sin\left(\frac{3\theta}{2}\right)}{\sin\left(\frac{\theta}{2}\right)}.$$

From the last expression it is easy to see that $\int f^2 d\mu = 1$. In analogy with the previous case, the second expression for f suggests choosing

$$\varphi_n(\theta) = \frac{\sin\left(\left(\frac{2n+1}{2}\right)\theta\right)}{\sin\left(\frac{\theta}{2}\right)}$$

for $n \geq 1$ (and $\varphi_0 = 1$), so that $\{\varphi_n\}_{n\geq 0}$ is orthonormal with respect to μ. Rewriting

$$\varphi_n(\theta) = 1 + \sum_{j=1}^{n}\left(e^{ij\theta} + e^{-ij\theta}\right)$$

shows that $\varphi_n(\theta)$ is indeed a monic polynomial of degree n in f. The kernel $K_N^{\mu}(x, y)$ as in Lemma 3.8 is then

$$K_N^{\mu}(x,y) = \sum_{j=0}^{N-1}\varphi_j(x)\overline{\varphi}_j(y) = 1 + \sum_{j=1}^{N-1}\frac{\sin\left(\left(\frac{2n+1}{2}\right)x\right)\sin\left(\left(\frac{2n+1}{2}\right)y\right)}{\sin\left(\frac{x}{2}\right)\sin\left(\frac{y}{2}\right)}.$$

The distribution of the eigenangles from the Weyl formula is

$$\begin{aligned}
&\frac{2^N}{N!\,\pi^N}\prod_{1\leq j\leq N}\sin^2\left(\frac{\theta_j}{2}\right)\prod_{1\leq j<k\leq N}(2\cos(\theta_k) - 2\cos(\theta_j))^2 d\theta_1\cdots d\theta_N \\
&= \frac{2^N}{N!\,\pi^N}\prod_{1\leq j\leq N}\sin^2\left(\frac{\theta_j}{2}\right)\left(\det\left[(1+2\cos(\theta_j))^{k-1}\right]_{j,k=1}^N\right)^2 d\theta_1\cdots d\theta_N \\
&= \frac{1}{N!}D_{N,N}d\mu(\theta_1)\cdots d\mu(\theta_N),
\end{aligned}$$

by part 1 of the lemma and definition of μ. Now, if $k \leq N$ and $A_1, \ldots, A_k \subseteq \Lambda$ are pairwise disjoint, then by the computation above and part 4 of the lemma,

$$\begin{aligned}
\mathbb{E}\left[\prod_{j=1}^{k} \mathcal{N}_{A_j}\right]
&= \frac{k!}{N!}\int_{\Lambda^N}\left(\sum_{1\leq j_1<\cdots<j_k\leq N}\prod_{\ell=1}^{k}\mathbb{1}_{A_\ell}(\theta_{j_\ell})\right) D_{N,N}(\theta_1,\ldots,\theta_N)d\mu(\theta_1)\cdots d\mu(\theta_N)\\
&= \int_{\Lambda^k}\left(\prod_{\ell=1}^{k}\mathbb{1}_{A_\ell}(\theta_\ell)\right) D_{k,N}(\theta_1,\ldots,\theta_k)d\mu(\theta_1)\cdots d\mu(\theta_k)\\
&= \frac{2^k}{\pi^k}\int_{A_1}\cdots\int_{A_k}\left(\det\left[1+\sum_{\ell=1}^{N-1}\frac{\sin\left(\left(\frac{2\ell+1}{2}\right)\theta_i\right)\sin\left(\left(\frac{2\ell+1}{2}\right)\theta_j\right)}{\sin\left(\frac{\theta_i}{2}\right)\sin\left(\frac{\theta_j}{2}\right)}\right]_{i,j=1}^{k}\right)\\
&\qquad\qquad\times\prod_{j=1}^{k}\sin^2\left(\frac{\theta_j}{2}\right)d\theta_j.
\end{aligned}$$

Now multiplying the ith row of matrix inside the determinant by one factor of $\sin\left(\frac{\theta_i}{2}\right)$ and the jth column by the other factor of $\sin\left(\frac{\theta_j}{2}\right)$, and each entry by 2 gives that this last expression is

$$\frac{1}{\pi^k}\int_{A_1}\cdots\int_{A_k}\left(\det\left[\sum_{\ell=0}^{N-1}\sin\left(\left(\frac{2\ell+1}{2}\right)\theta_i\right)\sin\left(\left(\frac{2\ell+1}{2}\right)\theta_j\right)\right]_{i,j=1}^{k}\right)\prod_{j=1}^{k}d\theta_j.$$

To treat the negative coset, one need only note that because we are in the odd case, the two cosets of the orthogonal group are related by multiplication by $-I$. The nontrivial eigenangles $(\psi_1, \ldots, \psi_N)$ of a Haar-distributed random matrix in $\mathbb{SO}^-(2N+1)$ are thus equal in distribution to $(\pi-\theta_1, \ldots, \pi-\theta_N)$, where $(\theta_1, \ldots, \theta_N)$ are the nontrivial eigenangles of a Haar-distributed random matrix in $\mathbb{SO}(2N+1)$, and the claimed formula then follows by changing variables.

Symplectic case: This is essentially the same as the previous case. Taking $d\mu = \frac{2}{\pi}\sin(\theta)d\theta, f(\theta) = 2\cos(\theta)$, and $\varphi_n(\theta) = \frac{\sin(n\theta)}{\sin(\theta)}$ for $n \geq 1$, the proof proceeds as above. □

For some purposes, the following alternative forms of the kernels can be more convenient. In all but the unitary case, they are the same functions; for the unitary case, the kernels are different functions but are unitarily similar and thus define the same point processes.

Proposition 3.9 *For $N \in \mathbb{N}$, let*

$$S_N(x) := \begin{cases} \sin\left(\frac{Nx}{2}\right) / \sin\left(\frac{x}{2}\right) & \text{if } x \neq 0, \\ N & \text{if } x = 0. \end{cases}$$

The nontrivial eigenvalue angles of uniformly distributed random matrices in any of $\mathbb{SO}(N)$, $\mathbb{SO}^-(N)$, $\mathbb{U}(N)$, $\mathbb{S}\mathbb{p}(2N)$ are a determinantal point process, with respect to uniform (probability) measure on Λ, with kernels as follows.

	$L_N(x,y)$	Λ
$\mathbb{U}(N)$	$S_N(x-y)$	$[0, 2\pi)$
$\mathbb{SO}(2N)$	$\frac{1}{2}\Big(S_{2N-1}(x-y) + S_{2N-1}(x+y)\Big)$	$[0, \pi)$
$\mathbb{SO}(2N+1)$	$\frac{1}{2}\Big(S_{2N}(x-y) - S_{2N}(x+y)\Big)$	$[0, \pi)$
$\mathbb{SO}^-(2N+1)$	$\frac{1}{2}\Big(S_{2N}(x-y) + S_{2N}(x+y)\Big)$	$[0, \pi)$
$\mathbb{S}\mathbb{p}(2N)$ $\mathbb{SO}^-(2N+2)$	$\frac{1}{2}\Big(S_{2N+1}(x-y) - S_{2N+1}(x+y)\Big)$	$[0, \pi)$

Exercise 3.10 Verify that these kernels define the same determinantal point processes as those given on page 71.

3.3 Matrix Moments

While we have in principle completely described the distribution of the eigenvalues of a random matrix by the Weyl density formula, in practice, using the Weyl density to answer natural questions about random matrices, particularly those having to do with asymptotic behavior for large matrices, can be rather difficult. Under such circumstances, one option is to consider the moments. The following remarkable moment formulae are due to Diaconis and Shahshahani [36]; the proofs give a nice illustration of the use of character theory in the study of random matrices from the classical compact groups. Because it is the simplest, we begin with the case of the unitary group.

Proposition 3.11 *Let U be a Haar-distributed random matrix in $\mathbb{U}(n)$ and let $Z_1, \ldots, Z_k$ be i.i.d. standard complex Gaussian random variables* *(i.e., complex-valued random variables whose real and imaginary parts are independent centered Gaussians with variance $\frac{1}{2}$).*

1. *Let $a = (a_1, \ldots, a_k)$ and $b = (b_1, \ldots, b_k)$ with $a_j, b_j \in \mathbb{N}$, and let $n \in \mathbb{N}$ be such that*

$$\max\left\{\sum_{j=1}^{k} ja_j, \sum_{j=1}^{k} jb_j\right\} \leq n.$$

Then

$$\mathbb{E}\left[\prod_{j=1}^{k}((\mathrm{Tr}(U^j))^{a_j}\overline{(\mathrm{Tr}(U^j))^{b_j}}\right] = \delta_{ab}\prod_{j=1}^{k} j^{a_j} a_j! = \mathbb{E}\left[\prod_{j=1}^{k}(\sqrt{j}Z_j)^{a_j}\overline{(\sqrt{j}Z_j)^{b_j}}\right].$$

2. *For any j, k, n, if U is Haar-distributed in $\mathbb{U}(n)$, then*

$$\mathbb{E}\left[\mathrm{Tr}(U^j)\overline{\mathrm{Tr}(U^k)}\right] = \delta_{jk}\min\{j, n\}.$$

Proof Let p_j denote the power sum symmetric function:

$$p_j(x_1, \ldots, x_n) = x_1^j + \cdots + x_n^j.$$

Then

$$\prod_{j=1}^{k}(\mathrm{Tr}(U^j))^{a_j} = \prod_{j=1}^{k} p_j^{a_j}(U),$$

where by abuse of notation, $p_j(U)$ denotes the scalar-valued function

$$p_j(U) = p_j(\lambda_1, \ldots, \lambda_n)$$

where $\lambda_1, \ldots, \lambda_n$ are the eigenvalues of U. In symmetric function theory, such a product of power sums (a so-called compound power sum) is denoted p_μ, where μ is the partition of the integer $K = a_1 + 2a_2 + \cdots + ka_k$ consisting of a_1 1's, a_2 2's, and so on. There are many different bases for the space of symmetric functions, one of which is the Schur functions, which also happen to be the irreducible characters of $\mathbb{U}(n)$ (see Section 1.3). Better still, the change-of-basis formula with which one can express compound power sums in terms of Schur functions is very explicit, with the coefficients given by irreducible characters of the symmetric group. Indeed, the conjugacy classes of the symmetric group are exactly described by cycle structure, and a cycle structure can be thought of as an integer partition (a_1 1-cycles, a_2 2-cycles, ..., a_k k-cycles corresponds to the partition of K above). Since the irreducible characters are in one-to-one correspondence with the conjugacy classes, this means that the irreducible characters of S_K can be indexed as $\{\chi_\lambda : \lambda \text{ a partition of } K\}$. It is possible to make explicit how a partition leads to an irreducible representation of S_K, but we will not go into this here. Instead, we will take as given the following change of basis formula, which is proved using the representation theory of the symmetric groups. For p_μ a compound power sum as above and $\{s_\lambda\}_{\lambda \vdash K}$ the Schur functions corresponding to partitions λ of the integer K,

$$p_\mu = \sum_{\lambda \vdash K} \chi_\lambda(\mu) s_\lambda.$$

Writing $L = b_1 + 2b_2 + \cdots + kb_k$ and taking ν to be the partition of L with b_1 1's, b_2 2's, etc.,

$$\begin{aligned}\mathbb{E}\left[\prod_{j=1}^{k} ((\mathrm{Tr}(U^j))^{a_j} \overline{(\mathrm{Tr}(U^j))^{b_j}}\right] &= \mathbb{E}\left[p_\mu(U)\overline{p_\nu(U)}\right] \\ &= \sum_{\lambda \vdash K, \pi \vdash L} \chi_\lambda(\mu)\overline{\chi_\pi(\nu)}\mathbb{E}\left[s_\lambda(U)\overline{s_\pi(U)}\right].\end{aligned}$$

Since $U \mapsto s_\lambda(U)$ are the irreducible characters of $\mathbb{U}(n)$, it follows from the first orthogonality relation (Proposition 1.12 in Section 1.3),

$$\mathbb{E}\left[s_\lambda(U)\overline{s_\pi(U)}\right] = \delta_{\lambda\pi} \mathbb{1}(\ell(\lambda) \leq n).$$

(Recall that $s_\lambda(U) = 0$ if $\ell(\lambda) > n$). The condition on n in the statement of the theorem is exactly that $\max\{K, L\} \leq n$, and so $\ell(\lambda) \leq n$ for each partition of K or L. That is,

$$\mathbb{E}\left[\prod_{j=1}^{k} ((\mathrm{Tr}(U^j))^{a_j} \overline{(\mathrm{Tr}(U^j))^{b_j}}\right] = \delta_{KL} \sum_{\lambda \vdash K} \chi_\lambda(\mu)\overline{\chi_\lambda(\nu)}.$$

Applying the second orthogonality relation (Proposition 1.13 of Section 1.3) to the irreducible characters of S_K gives that

$$\sum_{\lambda \vdash K} \chi_\lambda(\mu)\overline{\chi_\lambda(\nu)} = \delta_{\mu\nu} \frac{|S_K|}{c(\mu)},$$

where $c(\mu)$ is the size of the conjugacy class of μ in S_K. Since μ corresponds to permutations with a_1 1-cycles, a_2 2-cycles, and so on,

$$c(\mu) = \binom{K}{1^{a_1}, \ldots, k^{a_k}} \frac{(0!)^{a_1} \cdots ((k-1)!)^{a_k}}{a_1! \cdots a_k!},$$

where the multinomial coefficient $\binom{K}{1^{a_1}, \ldots, k^{a_k}}$ is the number of ways to divide the integers $1, \ldots, K$ into a_1 groups of size 1, a_2 groups of size 2, etc; the number of cyclic permutations of $\{i_1, \ldots, i_j\}$ is $(j-1)!$; and the factor of $a_1! \cdots a_k!$ in the denominator corresponds to orders in which one can write the cycles of a given length. We thus have that

$$\mathbb{E}\left[\prod_{j=1}^{k}((\mathrm{Tr}(U^j))^{a_j}\overline{(\mathrm{Tr}(U^j))^{b_j}}\right] = \delta_{KL}\delta_{\lambda\mu}\prod_{j=1}^{k} j^{a_j}a_j! = \delta_{ab}\prod_{j=1}^{k} j^{a_j}a_j!.$$

To see that this is the same as the mixed moment of complex Gaussian random variables as stated in the theorem, first note that a standard complex Gaussian random variable Z can be written as $Re^{i\theta}$ with $R^2 = Z_1^2 + Z_2^2$ for (Z_1, Z_2) independent real standard Gaussian variables, θ uniformly distributed in $[0, 2\pi)$, and (R, θ) independent. Moreover, $Z_1^2 + Z_2^2 \stackrel{d}{=} Y$, where Y is an exponential random variable with parameter 1. Then

$$\mathbb{E}[Z^a\overline{Z^b}] = \mathbb{E}\left[R^{a+b}\right]\mathbb{E}\left[e^{i(a-b)\theta}\right] = \delta_{ab}\mathbb{E}\left[Y^a\right] = \delta_{ab}a!.$$

To prove the second statement, by the argument above we have that

$$\mathbb{E}[\mathrm{Tr}(U^j)\overline{\mathrm{Tr}(U^k)}] = \delta_{jk}\sum_{\lambda\vdash j}|\chi_\lambda((j))|^2\mathbb{1}(\ell(\lambda)\le n),$$

where (j) is the trivial partition of j into the single part j. Evaluating the sum thus requires knowledge of the irreducible characters of the symmetric group. It is a fact (see exercise 4.16 in [47]) that $\chi_\lambda((j)) = 0$ unless λ is a so-called hook partition: a partition of the form $(a, 1, 1, \cdots, 1)$ (its Young diagram looks like a hook). In case that λ is a hook partition,

$$\chi_\lambda((j)) = (-1)^{\ell(\lambda)-1}.$$

There are $\min\{j, n\}$ hook partitions λ of j with $\ell(\lambda) \le n$, and so

$$\mathbb{E}[\mathrm{Tr}(U^j)\overline{\mathrm{Tr}(U^k)}] = \delta_{jk}\min\{j, n\}.$$

□

The orthogonal and symplectic cases are more complicated because the expansions of the power sum symmetric functions in terms of group characters are more difficult. The approach can nevertheless be extended to give the following.

Theorem 3.12 *Let U be a Haar-distributed random matrix in $\mathbb{O}(n)$, and let $Z_1, \ldots, Z_k$ be i.i.d. standard Gaussian random variables. Suppose that $a_1, \ldots, a_k$ are nonnegative integers such that $2\sum_{j=1}^{k} ja_j \le n$. Let η_j be 1 if j is even and 0 if j is odd. Then*

$$\mathbb{E}\left[\prod_{j=1}^{k}(\mathrm{Tr}(U^j))^{a_j}\right] = \prod_{j=1}^{k} f_j(a_j) = \mathbb{E}\left[\prod_{j=1}^{k}(\sqrt{j}Z_j + \eta_j)^{a_j}\right],$$

where

$$f_j(a) = \begin{cases} 0, & j, a \text{ odd}; \\ j^{\frac{a}{2}}(a-1)!!, & j \text{ odd}, a \text{ even}; \\ \sum_{s=0}^{\lfloor \frac{a}{2} \rfloor} \binom{a}{2s}(2s-1)!!j^s, & j \text{ even}. \end{cases}$$

Proof Let μ denote the partition of $K := a_1 + 2a_2 + \cdots + ka_k$ with a_1 1's, a_2 2's, and so on. Then as in the unitary case,

$$\prod_{j=1}^{k}(\mathrm{Tr}(U^j))^{a_j} = p_\mu(U),$$

the compound power symmetric function evaluated at the eigenvalues of U. As before, the key to the proof is to express p_μ in terms of the irreducible characters of the orthogonal group; the necessary expansion is rather less classical in this case and was developed in the early '90s by A. Ram.

Let V be the standard n-dimensional representation of $\mathbb{O}(n)$ (i.e., $U \in \mathbb{O}(n)$ acts by matrix-vector multiplication), and for $k \in \mathbb{N}$, recall that $V^{\otimes k}$ is a representation of $\mathbb{O}(n)$, where $U(v_1 \otimes \cdots \otimes v_k) = Uv_1 \otimes \cdots \otimes Uv_k$. The **Brauer algebra** $B_{k,n}$ is the algebra of linear transformations of $V^{\otimes k}$ that commute with the action of $\mathbb{O}(n)$. The Brauer algebra is not a group algebra, and so it does not fit into the usual framework of representation theory. Nevertheless, there is a notion of irreducible characters, and these play the role here that the irreducible characters of the symmetric group played in the case of random unitary matrices. Specifically, there is the following expansion of the compound power symmetric function:

$$p_\mu(x_1, \ldots, x_n) = \sum_{j=0}^{\lfloor \frac{K}{2} \rfloor} \sum_{\nu \vdash K-2j} \mathbb{1}_{\{\nu'_1 + \nu'_2 \leq n\}} \chi^{\nu}_{K,n}(\omega) t_\nu(x_1, \ldots, x_n),$$

where ν'_1, ν'_2 are the first two parts of the conjugate partition to ν (so that the indicator is nonzero if and only if there are at most n boxes in the first two columns of the Young diagram of ν); $\chi^{\nu}_{K,n}(\omega)$ is a character of $B_{K,n}$ corresponding to ν, being evaluated at an argument ω determined by μ (exactly how this works will be treated like a black box here); and t_ν is the character of $\mathbb{O}(n)$ corresponding to ν (see Theorem 1.17).

By the first orthogonality relation for characters (Proposition 1.12 in Section 1.3), $\mathbb{E}[t_\nu(U)] = 0$ unless t_ν is constant; this corresponds to the trivial partition $\nu = (0)$. It thus follows from the expansion formula that

$$\mathbb{E}\left[\prod_{j=1}^{k}(\operatorname{Tr}(U^j))^{a_j}\right] = \begin{cases} \chi_{K,n}^{(0)}(\omega), & K \text{ even;} \\ 0, & K \text{ odd.} \end{cases}$$

The formulae developed by Ram for evaluating these characters of the Brauer algebra give exactly the value for $\chi_{K,n}^{(0)}(\omega)$ quoted in the statement, as long as $n \geq 2K$. Seeing that this is the same as the mixed moment of Gaussians is immediate in the case that $\eta_j = 0$ (i.e., j is odd), and only requires expanding the binomial in the case $\eta_j = 1$. □

Brauer proved expansion formulae for the power sum symmetric functions in terms of the characters of the symplectic group as well; these also involve characters of the Brauer algebra as coefficients. The following theorem then follows by the same approach as in the orthogonal case. Recall that the eigenvalues of a symplectic matrix come in complex conjugate pairs, so traces of powers are necessarily real.

Theorem 3.13 *Let U be a Haar-distributed random matrix in $\mathbb{S}\mathbb{p}(2n)$, and let $Z_1, \ldots, Z_k$ be i.i.d. standard Gaussian random variables. Suppose that $a_1, \ldots, a_k$ are nonnegative integers such that $\sum_{j=1}^k ja_j \leq n$. Let η_j be 1 if j is even and 0 if j is odd, and let $f_j(a)$ be defined as in Theorem 3.12. Then*

$$\mathbb{E}\left[\prod_{j=1}^{k}(\operatorname{Tr}(U^j))^{a_j}\right] = \prod_{j=1}^{k}(-1)^{(j-1)a_j} f_j(a_j) = \mathbb{E}\left[\prod_{j=1}^{k}(\sqrt{j}Z_j - \eta_j)^{a_j}\right].$$

3.4 Patterns in Eigenvalues: Powers of Random Matrices

The structure of the eigenvalue distributions of random matrices from the classical compact groups is very rich, with intriguing patterns hidden somehow in the Weyl integration formula. One example is the following result of E. Rains.

Theorem 3.14 *Let $m \in \mathbb{N}$ be fixed and let $\widetilde{m} := \min\{m, N\}$. If $\sim$ denotes equality of eigenvalue distributions, then*

$$\mathbb{U}(N)^m \sim \bigoplus_{0 \leq j < \widetilde{m}} \mathbb{U}\left(\left\lceil \frac{N-j}{\widetilde{m}} \right\rceil\right)$$

That is, if U is a uniform $N \times N$ unitary matrix, the eigenvalues of U^m are distributed as those of $\widetilde{m}$ independent uniform unitary matrices of sizes

$$\left\lfloor \frac{N}{\widetilde{m}} \right\rfloor := \max\left\{k \in \mathbb{N} \mid k \leq \frac{N}{\widetilde{m}}\right\} \quad \text{and} \quad \left\lceil \frac{N}{\widetilde{m}} \right\rceil := \min\left\{k \in \mathbb{N} \mid k \geq \frac{N}{\widetilde{m}}\right\},$$

such that the sum of the sizes of the matrices is N.

In particular, if $m \geq N$, the eigenvalues of U^m are distributed exactly as N i.i.d. uniform points on $\mathbb{S}^1$.

For the proof, we will need the following preliminary lemma.

Lemma 3.15 *Let*

$$p(\theta_1, \dots, \theta_n) := \sum_{(p_1,\dots,p_n)} a_{(p_1,\dots,p_n)} \prod_{1 \leq k \leq n} e^{ip_k\theta_k}$$

be a Laurent polynomial in $\{e^{i\theta_k}\}_{k=1}^n$ that is a probability density with respect to $\frac{1}{(2\pi)^n} d\theta_1 \cdots d\theta_n$. If $(\Theta_1, \dots, \Theta_n)$ is distributed according to p, then the density of $(m\Theta_1, \dots, m\Theta_n)$ with respect to $\frac{1}{(2\pi)^n} d\theta_1 \cdots d\theta_n$ is

$$p^{(m)}(\theta_1, \dots, \theta_n) := \sum_{\substack{(p_1,\dots,p_n) \\ m|p_k \,\forall k}} a_{(p_1,\dots,p_n)} \prod_{1 \leq k \leq n} e^{i\left(\frac{p_k}{m}\right)\theta_k},$$

where $m\Theta_j$ is interpreted modulo 2π.

Proof To prove the lemma, it suffices to show that if $(\Phi_1, \dots, \Phi_n)$ is distributed according to $p^{(m)}$, then for any Laurent monomial $\mu(\theta_1, \dots, \theta_n) := \prod_{1 \leq k \leq n} e^{ir_k\theta_k}$,

$$\mathbb{E}\mu(m\Theta_1, \dots, m\Theta_n) = \mathbb{E}\mu(\Phi_1, \dots, \Phi_n).$$

Now, given two Laurent monomials

$$\mu_1(\theta_1, \dots, \theta_n) := \prod_{1 \leq k \leq n} e^{ir_k\theta_k} \qquad \mu_2(\theta_1, \dots, \theta_n) := \prod_{1 \leq k \leq n} e^{is_k\theta_k},$$

we have

$$\frac{1}{(2\pi)^n} \int_0^{2\pi} \cdots \int_0^{2\pi} \mu_1(\theta)\mu_2(\theta) d\theta_1 \cdots d\theta_n = \begin{cases} 1, & r_k + s_k = 0 \,\forall k; \\ 0, & \textit{otherwise}, \end{cases}$$

and so

$$\mathbb{E}\mu(m\Theta_1,\ldots,m\Theta_n) = \frac{1}{(2\pi)^n}\sum_{(p_1,\ldots,p_n)} a_{(p_1,\ldots,p_n)} \int_0^{2\pi}\cdots\int_0^{2\pi} \times \prod_{1\le k\le n} e^{i(p_k+mr_k)\theta_k}d\theta_1\ldots d\theta_n$$

$$= a_{(-mr_1,\ldots,-mr_n)}.$$

On the other hand,

$$\mathbb{E}\mu(\Phi_1,\ldots,\Phi_n) = \frac{1}{(2\pi)^n}\sum_{\substack{(p_1,\ldots,p_n)\\ m|p_k\,\forall k}} a_{(p_1,\ldots,p_n)} \times \int_0^{2\pi}\cdots\int_0^{2\pi}\prod_{1\le k\le n} e^{i(\frac{p_k}{m}+r_k)\theta_k}d\theta_1\ldots d\theta_n$$

$$= a_{(-mr_1,\ldots,-mr_n)},$$

which completes the proof. □

Proof of Theorem 3.14 For the proof, it is more convenient to index eigenvalues starting at 0 than at 1: the eigenvalues of U are denoted $\{e^{i\theta_k}\}_{0\le k<n}$.

By expanding the Vandermonde determinant as a sum over permutations, the Weyl density can be written as

$$\frac{1}{(2\pi)^n n!}\sum_{\sigma,\tau\in S_n}\operatorname{sgn}(\sigma\tau^{-1})\prod_{0\le k<n} e^{i(\sigma(k)-\tau(k))\theta_k}.$$

Then by Lemma 3.15, the density of the eigenvalues of U^m with respect to $\frac{1}{(2\pi)^n}d\theta_1\ldots d\theta_n$ is

$$\frac{1}{n!}\sum_{\substack{\sigma,\tau\in S_n,\\ m|(\sigma(k)-\tau(k))\,\forall k}}\operatorname{sgn}(\sigma\tau^{-1})\prod_{0\le k<n} e^{i\left(\frac{\sigma(k)-\tau(k)}{m}\right)\theta_k}.$$

Making the change of index $\ell=\tau(k)$ followed by the substitution $\pi=\sigma\circ\tau^{-1}$ reduces this expression to

$$\sum_{\substack{\pi\in S_n,\\ m|(\pi(\ell)-\ell)\,\forall\ell}}\operatorname{sgn}(\pi)\prod_{0\le\ell<n} e^{i\left(\frac{\pi(\ell)-\ell}{m}\right)\theta_\ell}.$$

Note that if $\pi\in S_n$ is such that $m\mid(\pi(\ell)-\ell)$ for all ℓ, then π permutes within conjugacy classes mod m, and can therefore be identified with the m

permutations it induces on those conjugacy classes. Specifically, for $0 \leq j < m$, define $\pi^{(j)} \in S_{\left\lceil \frac{n-j}{m} \right\rceil}$ by

$$\pi^{(j)}(k) = \frac{\pi(mk+j)-j}{m}.$$

(Note that $\mathrm{sgn}(\pi) = \prod_{0 \leq j < m} \mathrm{sgn}(\pi^{(j)})$.)

The density of the eigenvalues of U^m can thus be factored as

$$\sum_{\substack{\pi \in S_n, \\ m | (\pi(\ell)-\ell)\, \forall\, \ell}} \mathrm{sgn}(\pi) \prod_{0 \leq \ell < n} e^{i\left(\frac{\pi(\ell)-\ell}{m}\right)\theta_\ell}$$

$$= \sum_{\pi^{(0)} \in S_{\left\lceil \frac{n}{m} \right\rceil}} \cdots \sum_{\pi^{(m-1)} \in S_{\left\lceil \frac{n-m+1}{m} \right\rceil}} \prod_{0 \leq j < m} \left(\mathrm{sgn}(\pi^{(j)}) \prod_{0 \leq k < \left\lceil \frac{n-j}{m} \right\rceil} e^{i(\pi^{(j)}(k)-k)\theta_{km+j}} \right),$$

which is exactly the product of the eigenvalue densities of $U_0, \ldots, U_{m-1}$, where $\{U_j\}_{0 \leq j < m}$ are independent and U_j is distributed according to Haar measure on $\mathbb{O}\left(\left\lceil \frac{n-j}{m} \right\rceil\right)$. □

Notes and References

The Weyl integration formula is laid out in detail in Weyl's book [105]. It is discussed in varying detail and accessibility in many modern sources; I found the exposition of the unitary case in the notes [86] by Angela Pasquale particularly helpful.

The approach to the determinantal structure of the eigenvalue processes laid out in Section 3.2 follows that of chapter 5 of Katz and Sarnak's book [62]. Many other random matrix ensembles also have eigenvalue processes that are determinantal: the GUE, the complex Wishart matrices, and the complex Ginibre ensemble all have these properties (although the real counterparts do not), as does unitary Brownian motion.

The moments of traces of Haar-distributed random matrices have an interesting combinatorial connection; namely, they are related to the so-called increasing subsequence problem, which is about the distribution of the length of the longest increasing subsequence of a random permutation. (An increasing subsequence of a permutation π is a sequence $i_1 < i_2 < \cdots < i_k$ such that $\pi(i_1) < \pi(i_2) < \cdots < \pi(i_k)$.) In [89], Rains proved the following.

Theorem 3.16

1. *If U is distributed according to Haar measure on $\mathbb{U}(n)$ and $j \in \mathbb{N}$ is fixed, then $\mathbb{E}\left[|\operatorname{Tr}(U)|^{2j}\right]$ is equal to the number of permutations π of $\{1,\ldots,j\}$ such that π has no increasing subsequence of length greater than n.*
2. *If U is distributed according to Haar measure on $\mathbb{O}(n)$ and $j \in \mathbb{N}$ is fixed, then $\mathbb{E}\left[\operatorname{Tr}(U)^{j}\right]$ is equal to the number of permutations π of $\{1,\ldots,j\}$ such that $\pi^{-1} = \pi$, π has no fixed points, and π has no increasing subsequence of length greater than n.*
3. *If U is distributed according to Haar measure on $\mathbb{S}\mathbb{p}(2n)$ and $j \in \mathbb{N}$ is fixed, then $\mathbb{E}\left[\operatorname{Tr}(U)^{j}\right]$ is equal to the number of permutations π of $\{1,\ldots,j\}$ such that $\pi^{-1} = \pi$, π has no fixed points, and π has no decreasing subsequence of length greater than $2n$.*

The approach taken by Rains involved expansion in terms of characters, similar to the proof of Theorem 3.11. The connection between Haar-distributed random matrices and increasing subsequence problems was further explored by Baik and Rains [7], who showed that the moments of traces of random matrices from the classical groups are equal to the dimensions of certain invariant spaces of the groups (these subspaces having a natural connection to increasing subsequence problems).

Other approaches to Theorem 3.11 have since been developed: Stolz [98] gave an approach using invariant theory, and Hughes and Rudnick [57] gave an approach using cumulants, which in fact extended the range in which the matrix moments and the corresponding Gaussian moments match exactly.

In [90], Rains proved results analogous to Theorem 3.14 for random matrices in $\mathbb{SO}(n)$, $\mathbb{SO}^{-}(n)$, and $\mathbb{S}\mathbb{p}(2n)$; they are more complicated to state because of parity issues and because of the existence of trivial eigenvalues in some cases. In fact, Rains went much further, proving analogous results for general compact Lie groups.

4

Eigenvalue Distributions

Asymptotics

4.1 The Eigenvalue Counting Function

In this section we explore some of the consequences of the determinantal structure of the eigenvalue processes for the counting functions $\mathcal{N}_D$, and in particular for

$$\mathcal{N}_\theta = \#\{j : 0 \le \theta_j \le \theta\}.$$

We begin with some further background on determinantal point processes.

Some General Features of Determinantal Point Processes

The following remarkable property of certain determinantal point processes will be crucial in our analysis of the eigenvalue processes on the classical compact groups.

Theorem 4.1 *Let $\mathcal{X}$ be a determinantal point process on a compact metric measure space (Λ, μ) with kernel $K : \Lambda \times \Lambda \to \mathbb{C}$. Suppose that the corresponding integral operator $\mathcal{K} : L^2(\mu) \to L^2(\mu)$ defined by*

$$\mathcal{K}(f)(x) = \int K(x, y)f(y)\, d\mu(y) \tag{4.1}$$

is self-adjoint, nonnegative, and trace-class with eigenvalues in $[0, 1]$. *For $D \subseteq \Lambda$ measurable, let $K_D(x, y) = \mathbb{1}_D(x)K(x, y)\mathbb{1}_D(y)$ be the restriction of K to D, and denote by $\{\lambda_k\}_{k\in\mathcal{A}}$ the eigenvalues of the corresponding operator $\mathcal{K}_D$ on $L^2(D)$ ($\mathcal{A}$ may be finite or countable). Let $\mathcal{N}_D$ be the number of particles of the determinantal point process with kernel K which lie in D. Then*

$$\mathcal{N}_D \stackrel{d}{=} \sum_{k\in\mathcal{A}} \xi_k,$$

where "$\stackrel{d}{=}$" denotes equality in distribution and the ξ_k are independent Bernoulli random variables with $\mathbb{P}[\xi_k = 1] = \lambda_k$ and $\mathbb{P}[\xi_k = 0] = 1 - \lambda_k$.

Identifying the counting function of a point process as a sum of independent $\{0, 1\}$-valued random variables opens the door to the use of countless results of classical probability theory.

To prove the theorem, we need some preliminaries about the operator $\mathcal{K}$ defined in (4.1). We will assume in what follows that the kernel $K : \Lambda \times \Lambda \to \mathbb{C}$ is continuous, conjugate symmetric, and positive definite; i.e., for all $n \in \mathbb{N}$, $x_1, \ldots, x_n \in \Lambda$, and $z_1, \ldots, z_n \in \mathbb{C}$,

$$\sum_{i,j=1}^{n} K(x_i, x_j) z_i \overline{z_j} \geq 0.$$

We will also assume that that Λ is compact.

Under these assumptions, it is a classical result of operator theory that the operator $\mathcal{K} : L_2(\mu) \to L_2(\mu)$ is self-adjoint, nonnegative, and trace class; i.e., if $\{\lambda_j\}_{j\in J}$ are the eigenvalues of $\mathcal{K}$ (J may be finite or infinite), then $\sum_{j\in J} \lambda_j < \infty$. (The λ_j are necessarily positive, since $\mathcal{K}$ is nonnegative.) Moreover, $\mathcal{K}$ can be unitarily diagonalized; there are orthonormal eigenfunctions $\{\phi_j\}_{j\in J}$ of $\mathcal{K}$ such that, for any $f \in L_2(\mu)$,

$$\mathcal{K}(f) = \sum_{j\in J} \lambda_j \langle f, \phi_j \rangle \phi_j,$$

where the right-hand side converges in $L_2(\mu)$ for any $f \in L_2(\mu)$.

One can then conclude from a simple measure-theoretic argument that

$$K(x, y) = \sum_{j\in J} \lambda_j \phi_j(x) \overline{\phi}_j(y)$$

for $\mu \times \mu$ almost every (x, y).

If $\lambda_j \in \{0, 1\}$ for all j, then the operator $\mathcal{K}$ defined by K is just orthogonal projection in $L_2(\mu)$ onto the span of some of the ϕ_j; in this case, the following result shows that the total number of points in the process is deterministic and equal to the rank of $\mathcal{K}$.

Proposition 4.2 *Suppose that $\mathcal{X}$ is a determinantal point process on Λ with kernel*

$$K(x, y) = \sum_{j=1}^{N} \phi_j(x) \overline{\phi}_j(y),$$

where $\{\phi_j\}_{j=1}^{N}$ are orthonormal in $L_2(\mu)$. Then with probability one, $\mathcal{N}_\Lambda = N$.

Proof Observe first that if

$$K(x,y) = \sum_{j=1}^{N} \phi_j(x)\overline{\phi}_j(y),$$

then for any $n \in \mathbb{N}$ and

$$\big[K(x_i,x_j)\big]_{1\le i,j\le n} = \Phi\Phi^*,$$

where $\Phi = \big[\phi_j(x_i)\big]_{\substack{1\le i\le n\\ 1\le j\le N}} \in \mathrm{M}_{n,N}(\mathbb{C})$. In particular, if $n > N$, then $\det([K(x_i,x_j)]_{1\le i,j\le n}) = 0$, since $\Phi\Phi^*$ has rank at most N, and thus $\mathcal{N}_\Lambda \le N$. But also,

$$\mathbb{E}\mathcal{N}_\Lambda = \int_\Lambda K(x,x)d\mu(x) = \sum_{j=1}^{N} \int_\Lambda |\phi_j(x)|^2 d\mu(x) = N.$$

□

In the context of eigenvalue processes, we began with point processes and found them to be determinantal with explicitly described kernels. To prove Theorem 4.1, we will need to go in the other direction: namely, start with a kernel and from it get a point process. Recall that, by the results of Macchi and Soshnikov, this is always possible as long as the corresponding operator $\mathcal{K}$ is self-adjoint, trace class, and has all eigenvalues in $[0,1]$.

Proof of Theorem 4.1 First observe that it suffices to prove the Theorem when $D = \Lambda$, since if K defines a self-adjoint, nonnegative trace-class operator with eigenvalues in $[0,1]$, so does K_D.

Suppose that $\mathcal{K}$ is a finite-rank operator; i.e.,

$$K(x,y) = \sum_{j=1}^{N} \lambda_j \phi_j(x)\overline{\phi}_j(y)$$

for some finite N. Define a randomized version of K by

$$K_I(x,y) = \sum_{j=1}^{N} I_j \phi_j(x)\overline{\phi}_j(y),$$

where $\{I_j\}_{j=1}^N$ are independent Bernoulli random variables, with $\mathbb{P}[I_j = 1] = \lambda_j$ and $\mathbb{P}[I_j = 0] = 1 - \lambda_j$.

Let $\mathcal{X}_I$ denote the point process on Λ with kernel K_I; i.e., a random set of points constructed by first sampling the I_j to determine a kernel K_I, then sampling from the point process with kernel K_I. We will show that $\mathcal{X}_I$ has the same k-point correlation functions as $\mathcal{X}$, the point process defined by K.

Once we have shown this, the proof is completed as follows. Observe that K_I defines a projection operator. Specifically, $\mathcal{K}_I$ is orthogonal projection onto the span of those ϕ_j for which $I_j = 1$, and so the rank of $\mathcal{K}_I$ is $\sum_{j=1}^{N} I_j$. It thus follows from Proposition 4.2 that

$$|\mathcal{X}| \stackrel{d}{=} |\mathcal{X}_I| = \sum_{j=1}^{N} I_j.$$

We now show that the kernels K and K_I define the same point processes. Note that, as in the proof of Proposition 4.2 above, if $n \in \mathbb{N}$ and

$$C = \left[K(x_i, x_j)\right]_{1 \le i,j \le n} \qquad C_I = \left[K_I(x_i, x_j)\right]_{1 \le i,j \le n},$$

then $C = \Phi\, \mathbf{diag}(\lambda_1, \ldots, \lambda_N)\Phi^*$ and $C_I = \Phi\, \mathbf{diag}(I_1, \ldots, I_N)\Phi^*$, where $\Phi = \left[\phi_j(x_i)\right]_{\substack{1 \le i \le n \\ 1 \le j \le N}} \in \mathrm{M}_{n,N}(\mathbb{C})$. In particular, if $n > N$, then $\det(C) = \det(C_I) = 0$, since both matrices have rank at most N.

For $n \le N$, we must show that $\mathbb{E} \det(C_I) = \det(C)$, for which we use the Cauchy–Binet formula: for $C = AB$,

$$\det(C) = \sum_{1 \le k_1 < \cdots < k_n \le N} \det\left(A_{[n],\{k_1,\ldots,k_n\}}\right) \det\left(B_{\{k_1,\ldots,k_n\},[n]}\right),$$

where $[n] := \{1, \ldots, n\}$ and for sets of indices K_1, K_2, A_{K_1,K_2} denotes the matrix gotten from A by deleting all rows except those with indices in K_1 and all columns except those with indices in K_2. By multilinearity and independence,

$$\begin{aligned}
\mathbb{E} \det\left((\Phi\, \mathbf{diag}(I_1, \ldots, I_N))_{[n],\{k_1,\ldots,k_n\}}\right) &= \mathbb{E}\left[I_{k_1} \cdots I_{k_n} \det\left(\Phi_{[n],\{k_1,\ldots,k_n\}}\right)\right] \\
&= \lambda_{k_1} \cdots \lambda_{k_n} \det\left(\Phi_{[n],\{k_1,\ldots,k_n\}}\right) \\
&= \det\left((\Phi\, \mathbf{diag}(\lambda_1, \ldots, \lambda_N))_{[n],\{k_1,\ldots,k_n\}}\right).
\end{aligned}$$

The result now follows from the Cauchy–Binet formula.

If $K(x, y) = \sum_{j=1}^{\infty} \lambda_j \phi_j(x)\overline{\phi}_j(y)$ so that $\mathcal{K}$ is not a finite-rank operator, we consider the truncated kernel

$$K_N(x, y) = \sum_{j=1}^{N} \lambda_j \phi_j(x)\overline{\phi}_j(y)$$

corresponding to a rank N approximation of $\mathcal{K}$. Fix $n \in \mathbb{N}$; the argument above shows that for $x_1, \ldots, x_n \in \Lambda$,

$$\mathbb{E} \det\left[K_{N,I}(x_i, x_j)\right]_{1 \le i,j \le n} = \det\left[K_N(x_i, x_j)\right]_{1 \le i,j \le n},$$

where $K_{N,I}$ denotes the randomized version of K_N, with the λ_j replaced with i.i.d. Bernoulli random variables. By continuity of the determinant,

$$\det\left[K(x_i,x_j)\right]_{1\le i,j\le n} = \lim_{N\to\infty} \det\left[K_N(x_i,x_j)\right]_{1\le i,j\le n}$$

for all $x_1,\ldots,x_n \in \Lambda$ such that the sequences $\{\phi_j(x_i)\}_{j\ge 1}$ are square-summable (which is μ almost all of them). That is, we only need to show that

$$\mathbb{E}\left(\lim_{N\to\infty} \det\left[K_{N,I}(x_i,x_j)\right]_{1\le i,j\le n}\right) = \lim_{N\to\infty} \mathbb{E}\det\left[K_{N,I}(x_i,x_j)\right]_{1\le i,j\le n} \quad (4.2)$$

for μ almost every $x_1,\ldots,x_n$

By the Cauchy–Binet formula again,

$$\det\left[K_{N,I}(x_i,x_j)\right]_{1\le i,j\le n} = \sum_{1\le k_1<\cdots<k_n\le N} \det\left((\Phi D_I)_{[n],\{k_1,\ldots,k_n\}}\right)\det\left(\Phi^*_{\{k_1,\ldots,k_n\},[n]}\right),$$

where Φ is the half-infinite array with i-jth entry $\phi_j(x_i)$ (where $1\le i\le n$ and $j\ge 1$), and $D_I = \mathbf{diag}(I_1,I_2,\ldots)$. By multilinearity, this is

$$\sum_{1\le k_1<\cdots<k_n\le N} I_{k_1}\cdots I_{k_n}\left|\det\left(\Phi_{[n],\{k_1,\ldots,k_n\}}\right)\right|^2,$$

which is increasing in N for each choice of $I_1,I_2,\ldots$. The interchange of limit and expectation in (4.2) thus follows from the monotone convergence theorem.

The proof that the number of points in the process is a sum of independent Bernoulli random variables now follows as before: the fact that $\mathcal{K}$ is trace-class implies that with probability one, only finitely many I_j are nonzero. □

The following formulae are quite useful in analyzing the counting functions of determinantal point processes.

Lemma 4.3 *Let $K : \Lambda\times\Lambda \to \mathbb{C}$ be a continuous kernel such that the corresponding operator on $L^2(\mu)$ is a trace-class orthogonal projection. For $D\subseteq\Lambda$, denote by $\mathcal{N}_D$ the number of particles of the determinantal point process with kernel K which lie in D. Then*

$$\mathbb{E}\mathcal{N}_D = \int_D K(x,x)\,d\mu(x)$$

and

$$\operatorname{Var}\mathcal{N}_D = \int_D\int_{D^c} \left|K(x,y)\right|^2 d\mu(x)\,d\mu(y).$$

Proof The first statement is trivial. For the second, observe that if N is the (deterministic) total number of points in Λ

$$\mathbb{E}\left[\mathcal{N}_D^2\right] = \mathbb{E}\left[\mathcal{N}_D(N-\mathcal{N}_{D^c})\right] = N\mathbb{E}\mathcal{N}_D - \mathbb{E}\left[\mathcal{N}_D\mathcal{N}_{D^c}\right].$$

Now,

$$\begin{aligned}\mathbb{E}\left[\mathcal{N}_D\mathcal{N}_{D^c}\right] &= \int_D\int_{D^c} \rho_2(x,y)d\mu(x)d\mu(y)\\ &= \int_D\int_{D^c} \left[K(x,x)K(y,y) - K(x,y)K(y,x)\right]d\mu(x)d\mu(y)\\ &= \mathbb{E}\mathcal{N}_D\mathbb{E}\mathcal{N}_{D^c} - \int_D\int_{D^c} |K(x,y)|^2 d\mu(x)d\mu(y),\end{aligned}$$

using the fact that $K(y,x) = \overline{K(x,y)}$. Making use again of the relation $\mathcal{N}_{D^c} = N - \mathcal{N}_D$, we have from above that

$$\mathbb{E}\left[\mathcal{N}_D^2\right] = (\mathbb{E}\mathcal{N}_D)^2 + \int_D\int_{D^c} |K(x,y)|^2 d\mu(x)d\mu(y),$$

and the variance formula follows. □

Exercise 4.4 Confirm that the kernels K_n and L_n of Propositions 3.7 and 3.9 satisfy the conditions of the lemma.

The following lemma, whose proof we omit, relates the limiting behavior of the counting functions of a sequence of determinantal point processes on $\mathbb{R}$ to the counting functions of a limiting process, when there is convergence of the corresponding kernels.

Lemma 4.5 *Let $\{K_N(x,y)\}_{N\in\mathbb{N}}$ be a sequence of kernels of determinantal point processes, and suppose that there is a kernel $K(x,y)$ such that $K_N(x,y) \to K(x,y)$ uniformly on compact sets. Let $\mathcal{N}^{(N)}(\cdot)$ denote the counting function of the process with kernel K_N and $\mathcal{N}(\cdot)$ the counting function of the process with kernel K. Let $m \in \mathbb{N}$ and let $\{D_\ell\}_{\ell=1}^m$ be a finite collection of compact disjoint subsets of $\mathbb{R}$. Then the random vector $(\mathcal{N}^{(N)}(D_1), \ldots, \mathcal{N}^{(N)}(D_m))$ converges in distribution to the random vector $(\mathcal{N}(D_1), \ldots, \mathcal{N}(D_m))$.*

For a proof, see [1, section 4.2.8].

Asymptotics for the Eigenvalue Counting Function

We now return to the counting functions of the eigenvalue processes on the classical compact groups. Consider the *sine kernel*

$$K(x,y) = \frac{\sin(\pi(x-y))}{\pi(x-y)}$$

on $\mathbb{R}$. The sine kernel is the kernel, with respect to Lebesgue measure, of an unbounded determinantal point process on $\mathbb{R}$ called the sine kernel process. The most classical result on the asymptotics of the eigenvalue counts is that, when suitably rescaled, they converge to the sine kernel process as the matrix size tends to infinity, as follows.

Theorem 4.6 *Let $\{x_1, \ldots, x_N\}$ be the nontrivial eigenangles of a random matrix distributed according to Haar measure in one of $\mathbb{U}(N)$, $\mathbb{SO}(2N)$, $\mathbb{SO}^-(2N+2)$, $\mathbb{SO}(2N+1)$, $\mathbb{SO}^-(2N+1)$, or $\mathbb{S}\mathrm{p}(2N)$, recentered and rescaled to lie in $\left[-\frac{N}{2}, \frac{N}{2}\right]$. For a compact set $D \subseteq \mathbb{R}$, let*

$$\mathcal{N}^{(N)}(D) := \#\{j : x_j \in D\}.$$

Let χ denote the sine kernel process on $\mathbb{R}$ and let $\mathcal{S}(D)$ denote the number of points of χ in D.

For any collection $\{D_\ell\}_{\ell=1}^m$ of compact disjoint subsets of $\mathbb{R}$, the random vector $(\mathcal{N}^{(N)}(D_1), \ldots, \mathcal{N}^{(N)}(D_m))$ converges in distribution to the random vector $(\mathcal{S}(D_1), \ldots, \mathcal{S}(D_m))$.

Proof In each case, this is an easy consequence of Lemma 4.5. First consider the unitary case: let $\theta_1, \ldots, \theta_N$ be the angles of the eigenvalues of a random unitary matrix. By a change of variables, the kernel for the recentered, rescaled process $\left\{\frac{N\theta_1}{2\pi} - \frac{N}{2}, \ldots, \frac{N\theta_N}{2\pi} - \frac{N}{2}\right\}$ on $\left[-\frac{N}{2}, \frac{N}{2}\right]$ is given by

$$\frac{1}{N}\frac{\sin(\pi(x-y))}{\sin\left(\frac{\pi(x-y)}{N}\right)} \longrightarrow \frac{\sin(\pi(x-y))}{\pi(x-y)}$$

as $N \to \infty$, uniformly on compact sets.

Next consider the case of $\mathbb{SO}(2N)$. Let $\{\theta_1, \ldots, \theta_N\} \subseteq [0, \pi]$ denote the nontrivial eigenangles. Again by changing variables, the kernel for the recentered, rescaled process $\left\{\frac{N\theta_1}{\pi} - \frac{N}{2}, \ldots, \frac{N\theta_N}{\pi} - \frac{N}{2}\right\}$ on $\left[-\frac{N}{2}, \frac{N}{2}\right]$ is

$$\begin{aligned}
&\frac{1}{N}L_N^{\mathbb{SO}(N)}\left(\frac{\pi}{N}x + \frac{\pi}{2}, \frac{\pi}{N}x + \frac{\pi}{2}\right) \\
&\quad = \frac{\sin\left(\left(1 - \frac{1}{2N}\right)\pi(x-y)\right)}{2N\sin\left(\frac{\pi(x-y)}{2N}\right)} + \frac{\sin\left(\left(1 - \frac{1}{2N}\right)\pi(x+y) + N\pi - \frac{\pi}{2}\right)}{2N\sin\left(\frac{\pi(x+y)}{2N} + \frac{\pi}{2}\right)} \\
&\quad = \frac{\sin\left(\left(1 - \frac{1}{2N}\right)\pi(x-y)\right)}{2N\sin\left(\frac{\pi(x-y)}{2N}\right)} - \frac{\cos\left(\left(1 - \frac{1}{2N}\right)\pi(x+y) + N\pi\right)}{2N\cos\left(\frac{\pi(x+y)}{2N}\right)},
\end{aligned}$$

which again tends to the sine kernel uniformly on compact sets.

The remaining cases are essentially identical. □

We now specialize to the counting functions for arcs:

$$\mathcal{N}_\theta = \#\{j : 0 \le \theta_j \le \theta\},$$

where as usual θ varies over $[0, 2\pi]$ in the unitary case and $\theta \in [0, \pi]$ in all other cases.

We begin by using Lemma 4.3 to calculate means and variances.

Proposition 4.7 *1. Let U be uniform in $\mathbb{U}(N)$. For $\theta \in [0, 2\pi)$, let $\mathcal{N}_\theta$ be the number of eigenvalues angles of U in $[0, \theta)$. Then*

$$\mathbb{E}\mathcal{N}_\theta = \frac{N\theta}{2\pi}.$$

2. Let U be uniform in one of $\mathbb{SO}(2N)$, $\mathbb{SO}^-(2N+2)$, $\mathbb{SO}(2N+1)$, $\mathbb{SO}^-(2N+1)$, or $\mathbb{S}\mathbb{p}(2N)$. For $\theta \in [0, \pi)$, let $\mathcal{N}_\theta$ be the number of nontrivial eigenvalue angles of U in $[0, \theta)$. Then

$$\left|\mathbb{E}\mathcal{N}_\theta - \frac{N\theta}{\pi}\right| < 1.$$

Proof The equality for the unitary group follows from symmetry considerations, or immediately from Proposition 3.9 and Lemma 4.3.

In the case of $\mathbb{S}\mathbb{p}(2N)$ or $\mathbb{SO}^-(2N+2)$, by Proposition 3.7 and Lemma 4.3,

$$\mathbb{E}\mathcal{N}_\theta = \frac{1}{\pi}\int_0^\theta \sum_{j=1}^N 2\sin^2(jx)\,dx = \frac{N\theta}{\pi} - \frac{1}{2\pi}\sum_{j=1}^N \frac{\sin(2j\theta)}{j}.$$

Define $a_0 = 0$ and $a_j = \sum_{k=1}^j \sin(2k\theta)$, and let $b_j = \frac{1}{j+1}$. Then by summation by parts,

$$\sum_{j=1}^N \frac{\sin(2j\theta)}{j} = \frac{a_N}{N} + \sum_{j=1}^{N-1} \frac{a_j}{j(j+1)}.$$

Trivially, $|a_N| \le N$. Now observe that

$$\begin{aligned} a_j &= \operatorname{Im}\left[\sum_{k=1}^j e^{2ik\theta}\right] = \operatorname{Im}\left[e^{2i\theta}\frac{e^{2ij\theta}-1}{e^{2i\theta-1}}\right] \\ &= \operatorname{Im}\left[e^{i(j+1)\theta}\frac{\sin(j\theta)}{\sin(\theta)}\right] = \frac{\sin((j+1)\theta)\sin(j\theta)}{\sin(\theta)}. \end{aligned}$$

Since $|a_j|$ is invariant under the substitution $\theta \mapsto \pi - \theta$, it suffices to assume that $0 < \theta \le \pi/2$. In that case $\sin(\theta) \ge 2\theta/\pi$, and so

$$\sum_{j=1}^{N-1} \frac{|a_j|}{j(j+1)} \leq \frac{\pi}{2\theta} \left[\sum_{1 \leq j \leq 1/\theta} \theta^2 + \sum_{1/\theta < j \leq N-1} \frac{1}{j(j+1)} \right].$$

Clearly, $\sum_{1 \leq j \leq 1/\theta} \theta^2 \leq \theta$. For the second term, note that

$$\sum_{1/\theta < j \leq N-1} \frac{1}{j(j+1)} = \sum_{1/\theta < j \leq N-1} \left(\frac{1}{j} - \frac{1}{j+1} \right) = \frac{1}{j_0} - \frac{1}{N},$$

where j_0 denotes the first index in the sum. That is,

$$\sum_{j=1}^{N-1} \frac{|a_j|}{j(j+1)} \leq \frac{\pi}{2\theta} (\theta + \theta) = \pi.$$

All together,

$$\left| \mathbb{E}\mathcal{N}_\theta - \frac{N\theta}{\pi} \right| \leq \frac{1+\pi}{2\pi}.$$

The other cases are handled similarly. □

Proposition 4.8 *Let U be uniform in one of $\mathbb{U}(N)$, $\mathbb{SO}(2N)$, $\mathbb{SO}^-(2N+2)$, $\mathbb{SO}(2N+1)$, $\mathbb{SO}^-(2N+1)$, or $\mathbb{S}\mathbb{p}(2N)$. For $\theta \in [0, 2\pi)$ (in the unitary case) or $\theta \in [0, \pi]$ (in all other cases), let $\mathcal{N}_\theta$ be the number of eigenvalue angles of U in $[0, \theta)$. For each group or coset, there is a constant c depending only on the group or coset, such that*

$$\operatorname{Var} \mathcal{N}_\theta \leq c (\log N + 1).$$

Proof We treat the unitary case, which is the simplest, first. Note that if $\theta \in (\pi, 2\pi)$, then $\mathcal{N}_\theta \stackrel{d}{=} N - \mathcal{N}_{2\pi-\theta}$, and so it suffices to assume that $\theta \leq \pi$. By Proposition 3.9 and Lemma 4.3,

$$\begin{aligned}
\operatorname{Var} \mathcal{N}_\theta &= \frac{1}{4\pi^2} \int_0^\theta \int_\theta^{2\pi} S_N(x-y)^2 \, dx \, dy \\
&= \frac{1}{4\pi^2} \int_0^\theta \int_{\theta-y}^{2\pi-y} \frac{\sin^2\left(\frac{Nz}{2}\right)}{\sin^2\left(\frac{z}{2}\right)} \, dz \, dy \\
&= \frac{1}{4\pi^2} \left[\int_0^\theta \frac{z \sin^2\left(\frac{Nz}{2}\right)}{\sin^2\left(\frac{z}{2}\right)} \, dz + \int_\theta^{2\pi-\theta} \frac{\theta \sin^2\left(\frac{Nz}{2}\right)}{\sin^2\left(\frac{z}{2}\right)} \, dz \right. \\
&\qquad \left. + \int_{2\pi-\theta}^{2\pi} \frac{(2\pi - z) \sin^2\left(\frac{Nz}{2}\right)}{\sin^2\left(\frac{z}{2}\right)} \, dz \right] \\
&= \frac{1}{2\pi^2} \left[\int_0^\theta \frac{z \sin^2\left(\frac{Nz}{2}\right)}{\sin^2\left(\frac{z}{2}\right)} \, dz + \int_\theta^\pi \frac{\theta \sin^2\left(\frac{Nz}{2}\right)}{\sin^2\left(\frac{z}{2}\right)} \, dz \right],
\end{aligned}$$

where the third equality follows by changing the order of integration and evaluating the resulting inner integrals. For the first integral, since $\sin\left(\frac{z}{2}\right) \geq \frac{z}{\pi}$ for all $z \in [0, \theta]$, if $\theta > \frac{1}{N}$, then

$$\int_0^\theta \frac{z \sin^2\left(\frac{Nz}{2}\right)}{\sin^2\left(\frac{z}{2}\right)} dz \leq \int_0^{\frac{1}{N}} \frac{(\pi N)^2 z}{4} dz + \int_{\frac{1}{N}}^\theta \frac{\pi^2}{z} dz = \pi^2\left(\frac{1}{8} + \log(N) + \log(\theta)\right).$$

If $\theta \leq \frac{1}{N}$, there is no need to break up the integral, and one simply has the bound $\frac{(\pi N\theta)^2}{8} \leq \frac{\pi^2}{8}$. Similarly, if $\theta < \frac{1}{N}$, then

$$\begin{aligned}\int_\theta^\pi \frac{\theta \sin^2\left(\frac{Nz}{2}\right)}{\sin^2\left(\frac{z}{2}\right)} dz &\leq \int_\theta^{\frac{1}{N}} \frac{\theta(\pi N)^2}{4} dz + \int_{\frac{1}{N}}^\pi \frac{\pi^2\theta}{z^2} dz \\ &= \frac{\pi^2\theta N}{4}(1 - N\theta) + \pi^2 N\theta - \pi\theta \leq \frac{5\pi^2}{4};\end{aligned}$$

if $\theta \geq \frac{1}{N}$, there is no need to break up the integral, and one simply has a bound of π^2.

All together,

$$\operatorname{Var} \mathcal{N}_\theta \leq \log(N) + \frac{11}{16}.$$

The remaining cases are similar but more complicated, since the remaining kernels from Proposition 3.9 are sums of two terms. We will sketch the proof for $\mathbb{SO}(2N)$ and leave filling in the details and treating the remaining cases as exercises for the extremely dedicated reader.

Let $\theta \in [0, \pi]$; by Proposition 3.9 and Lemma 4.3,

$$\begin{aligned}\operatorname{Var} \mathcal{N}_\theta &= \frac{1}{4\pi^2} \int_0^\theta \int_\theta^\pi \left[S_{2N-1}(x - y) + S_{2N-1}(x + y)\right]^2 dx dy \\ &= \frac{1}{4\pi^2} \left\{ \int_0^\theta \int_\theta^\pi \left[S_{2N-1}(x - y)\right]^2 dx dy \right. \\ &\quad + 2\int_0^\theta \int_\theta^\pi S_{2N-1}(x - y) S_{2N-1}(x + y) dx dy \\ &\quad \left. + \int_0^\theta \int_\theta^\pi \left[S_{2N-1}(x + y)\right]^2 dx dy \right\}.\end{aligned}$$

Note that the integrals are invariant under the simultaneous substitutions $s = \pi - x$ and $t = \pi - y$, and so we may assume that $\theta \in \left[0, \frac{\pi}{2}\right]$.

Now, the first term is essentially identical to the unitary case and is bounded in the same way by $\frac{13}{32} + \frac{1}{4}\log(2N - 1)$.

The final term is similar but easier: one simply lets $z = x + y$, changes the order of integration, and bounds the resulting integrals as in the unitary case.

The numerator can be bounded by 1 in all cases, and the resulting bound on the third term is a constant, independent of N.

For the cross term, making the substitutions $z = x + y$ and $w = x - y$ yields

$$\begin{aligned}
&\int_0^\theta \int_\theta^\pi S_{2N-1}(x-y)S_{2N-1}(x+y)dxdy \\
&= \int_\theta^{2\theta} \int_{2\theta-z}^{z} \frac{\sin\left(\left(\frac{2N-1}{2}\right)z\right)\sin\left(\left(\frac{2N-1}{2}\right)w\right)}{\sin\left(\frac{z}{2}\right)\sin\left(\frac{w}{2}\right)} dwdz \\
&+ \int_{2\theta}^{\pi} \int_{z-2\theta}^{z} \frac{\sin\left(\left(\frac{2N-1}{2}\right)z\right)\sin\left(\left(\frac{2N-1}{2}\right)w\right)}{\sin\left(\frac{z}{2}\right)\sin\left(\frac{w}{2}\right)} dwdz \qquad (4.3)\\
&+ \int_{\pi}^{\pi+\theta} \int_{z-2\theta}^{2\pi-z} \frac{\sin\left(\left(\frac{2N-1}{2}\right)z\right)\sin\left(\left(\frac{2N-1}{2}\right)w\right)}{\sin\left(\frac{z}{2}\right)\sin\left(\frac{w}{2}\right)} dwdz.
\end{aligned}$$

To handle the first summand, first observe that $w \leq z \leq 2\theta \leq \pi$, and so

$$\sin\left(\frac{z}{2}\right)\sin\left(\frac{w}{2}\right) \geq \frac{zw}{\pi^2}.$$

Evaluating the resulting inner integral, if $\frac{1}{2N-1} \leq 2\theta - z$, estimating the numerator by 1 yields

$$\int_{2\theta-z}^{z} \frac{\sin\left(\left(\frac{2N-1}{2}\right)w\right)}{w} dw \leq \log(z) - \log(2\theta - z) \leq \log(\pi) + \log(2N-1).$$

If $2\theta - z \leq \frac{1}{2N-1} \leq z$, then using the estimate $\sin(x) \leq x$ in the first part of the interval and $\sin(x) \leq 1$ in the second part yields

$$\begin{aligned}
\int_{2\theta-z}^{z} \frac{\sin\left(\left(\frac{2N-1}{2}\right)w\right)}{w} dw &\leq \int_{2\theta-z}^{\frac{1}{2N-1}} \left(\frac{2N-1}{2}\right) dw + \int_{\frac{1}{2N-1}}^{z} \frac{1}{w} dw \\
&\leq \frac{1}{2} + \log(z) + \log(2N-1) \\
&\leq \frac{1}{2} + \log(\pi) + \log(2N-1).
\end{aligned}$$

Finally, if $z \leq \frac{1}{2N-1}$, then using the estimate $\sin(x) \leq x$ yields

$$\int_{2\theta-z}^{z} \frac{\sin\left(\left(\frac{2N-1}{2}\right)w\right)}{w} dw \leq \int_{2\theta-z}^{z} \left(\frac{2N-1}{2}\right) dw = \left(\frac{2N-1}{2}\right)(2z - 2\theta) \leq 1.$$

The first term of (4.3) is thus bounded by

$$\pi^2 \int_\theta^{2\theta} \left[\frac{1}{2} + \log(\pi) + \log(2N-1)\right] \frac{\sin\left(\left(\frac{2N-1}{2}\right)z\right)}{z} dz$$
$$\leq \pi^2 \left[\frac{1}{2} + \log(\pi) + \log(2N-1)\right] \int_\theta^{2\theta} \frac{1}{z} dz$$
$$= \pi^2 \left[\frac{1}{2} + \log(\pi) + \log(2N-1)\right] \log(2).$$

For the second term, if $\frac{1}{2N-1} \leq z - 2\theta$, then the inner integral is

$$\int_{z-2\theta}^{z} \frac{\sin\left(\left(\frac{2N-1}{2}\right)w\right)}{w} dw \leq \int_{z-2\theta}^{z} \frac{1}{w} dw = \log(z) - \log(z-2\theta)$$
$$= \log\left(1 + \frac{2\theta}{z-2\theta}\right).$$

The right-most expression is a decreasing function of z, and so it is maximized in this regime for $z = \frac{1}{2N-1} + 2\theta$; that is, if $\frac{1}{2N-1} \leq z - 2\theta$, then

$$\int_{z-2\theta}^{z} \frac{\sin\left(\left(\frac{2N-1}{2}\right)w\right)}{w} dw \leq \log\left(\frac{1}{2N-1} + 2\theta\right) + \log(2N-1)$$
$$= \log(1 + 2\theta(2N-1)).$$

If $z - 2\theta \leq \frac{1}{2N-1} \leq z$, then breaking up the integral as above yields

$$\int_{z-2\theta}^{z} \frac{\sin\left(\left(\frac{2N-1}{2}\right)w\right)}{w} dw \leq \int_{z-2\theta}^{\frac{1}{2N-1}} \left(\frac{2N-1}{2}\right) dw + \int_{\frac{1}{2N-1}}^{z} \frac{1}{w} dw$$
$$\leq \frac{1}{2} + \log(z) + \log(2N-1) = \frac{1}{2} + \log\big(z(2N-1)\big).$$

If $z \leq \frac{1}{2N-1}$, then

$$\int_{z-2\theta}^{z} \frac{\sin\left(\left(\frac{2N-1}{2}\right)w\right)}{w} dw \leq \int_{z-2\theta}^{z} \left(\frac{2N-1}{2}\right) dw = 2\theta\left(\frac{2N-1}{2}\right)$$
$$\leq 2z\left(\frac{2N-1}{2}\right) \leq 1.$$

If $2\theta < \frac{1}{2N-1}$, then the second term of (4.3) is bounded using all three estimates above:

$$\begin{aligned}
&\int_{2\theta}^{\pi}\int_{z-2\theta}^{z} \frac{\sin\left(\left(\frac{2N-1}{2}\right)z\right)\sin\left(\left(\frac{2N-1}{2}\right)w\right)}{\sin\left(\frac{z}{2}\right)\sin\left(\frac{w}{2}\right)}dwdz \\
&\quad \leq \pi^2 \int_{2\theta}^{\frac{1}{2N-1}} \frac{\sin\left(\left(\frac{2N-1}{2}\right)z\right)}{z}dz \\
&\quad + \pi^2 \int_{\frac{1}{2N-1}}^{\frac{1}{2N-1}+2\theta}\left[\frac{1}{2}+\log\left(z(2N-1)\right)\right]\frac{\sin\left(\left(\frac{2N-1}{2}\right)z\right)}{z}dz \\
&\quad + \pi^2\log(2\theta(2N-1)+1)\int_{\frac{1}{2N-1}+2\theta}^{\pi}\frac{\sin\left(\left(\frac{2N-1}{2}\right)z\right)}{z}dz.
\end{aligned}$$

Now,

$$\int_{2\theta}^{\frac{1}{2N-1}} \frac{\sin\left(\left(\frac{2N-1}{2}\right)z\right)}{z}dz \leq \left(\frac{2N-1}{2}\right)\left(\frac{1}{2N-1}-2\theta\right) \leq \frac{1}{2},$$

and

$$\begin{aligned}
&\int_{\frac{1}{2N-1}}^{\frac{1}{2N-1}+2\theta}\left[\frac{1}{2}+\log\left(z(2N-1)\right)\right]\frac{\sin\left(\left(\frac{2N-1}{2}\right)z\right)}{z}dz \\
&\quad \leq \left(\frac{2N-1}{2}\right)\theta + \int_{1}^{1+2\theta(2N-1)}\frac{\log(s)}{s}ds \leq \frac{1}{4}+\frac{\log^2(1+2\theta(2N-1))}{2} \\
&\quad \leq \frac{1}{4}+\frac{\log^2(2)}{2}.
\end{aligned}$$

Finally,

$$\begin{aligned}
&\log(2\theta(2N-1)+1)\int_{\frac{1}{2N-1}+2\theta}^{\pi}\frac{\sin\left(\left(\frac{2N-1}{2}\right)z\right)}{z}dz \leq \log(2)\int_{\frac{1}{2N-1}+2\theta}^{\pi}\frac{1}{z}dz \\
&\quad \leq \log(2)\big[\log(\pi)+\log(2N-1)\big],
\end{aligned}$$

and so in the case that $2\theta < \frac{1}{2N-1}$, the second term of (4.3) is bounded by $c\log(2N-1)$.

If $2\theta \geq \frac{1}{2N-1}$, then from the estimates above,

$$\int_{2\theta}^{\pi}\int_{z-2\theta}^{z}\frac{\sin\left(\left(\frac{2N-1}{2}\right)z\right)\sin\left(\left(\frac{2N-1}{2}\right)w\right)}{\sin\left(\frac{z}{2}\right)\sin\left(\frac{w}{2}\right)}dwdz$$
$$\leq \pi^2\int_{2\theta}^{2\theta+\frac{1}{2N-1}}\left[\frac{1}{2}+\log(z(2N-1))\right]\frac{1}{z}dz+\int_{2\theta+\frac{1}{2N-1}}^{\pi}\log\left(1+\frac{2\theta}{z-2\theta}\right)\frac{1}{z}dz.$$

For the first term,

$$\int_{2\theta}^{2\theta+\frac{1}{2N-1}}\left[\frac{1}{2}+\log(z(2N-1))\right]\frac{1}{z}dz$$
$$\leq\left[\frac{1}{2}+\log(2\theta(2N-1)+1)\right]\log\left(1+\frac{1}{\theta(2N-1)}\right)$$
$$\leq\left[\frac{1}{2}+\log(\pi(2N-1)+1)\right]\log(3).$$

For the second term, using the fact that $\log\left(1+\frac{2\theta}{z-2\theta}\right)$ is decreasing as a function of z,

$$\int_{2\theta+\frac{1}{2N-1}}^{\pi}\log\left(1+\frac{2\theta}{z-2\theta}\right)\frac{1}{z}dz$$
$$\leq\log\left(1+2\theta(2N-1)\right)\int_{2\theta+\frac{1}{2N-1}}^{2\left(2\theta+\frac{1}{2N-1}\right)}\frac{1}{z}dz$$
$$+\log\left(1+\frac{2\theta}{2\theta+\frac{1}{2N-1}}\right)\int_{2\left(2\theta+\frac{1}{2N-1}\right)}^{\pi}\frac{1}{z}dz$$
$$\leq\log\left(1+2\theta(2N-1)\right)\log(2)+\log(2)\left[\log(\pi)-\log\left(4\theta+\frac{2}{2N-1}\right)\right]$$
$$=\log(2)\log\left(\frac{\pi}{2}\right)+\log(2)\log(2N-1),$$

completing the bound of the second term of (4.3) in the case $2\theta\geq\frac{1}{2N-1}$.

Finally, the third term of (4.3) is

$$\int_{\pi}^{\pi+\theta}\int_{z-2\theta}^{2\pi-z}\frac{\sin\left(\left(\frac{2N-1}{2}\right)z\right)\sin\left(\left(\frac{2N-1}{2}\right)w\right)}{\sin\left(\frac{z}{2}\right)\sin\left(\frac{w}{2}\right)}dwdz$$
$$\leq\sqrt{2}\pi\int_{\pi}^{\pi+\theta}\int_{z-2\theta}^{2\pi-z}\frac{\sin\left(\left(\frac{2N-1}{2}\right)w\right)}{w}dwdz,$$

since $\frac{z}{2}\in\left[\frac{\pi}{2},\frac{3\pi}{4}\right]$ and thus $\sin\left(\frac{z}{2}\right)\geq\frac{1}{\sqrt{2}}$.

Now, if $\frac{1}{2N-1} \leq z - 2\theta$, then

$$\int_{z-2\theta}^{2\pi - z} \frac{\sin\left(\left(\frac{2N-1}{2}\right) w\right)}{w} dw \leq \log(2\pi - z) - \log(z - 2\theta)$$
$$\leq \log(2\pi) + \log(2N-1),$$

and if $z - 2\theta \leq \frac{1}{2N-1} \leq 2\pi - z$, then

$$\int_{z-2\theta}^{2\pi - z} \frac{\sin\left(\left(\frac{2N-1}{2}\right) w\right)}{w} dw \leq \int_{z-2\theta}^{\frac{1}{2N-1}} \left(\frac{2N-1}{2}\right) dw + \int_{\frac{1}{2N-1}}^{2\pi - z} \frac{1}{w} dw$$
$$\leq \frac{1}{2} + \log(2\pi) + \log(2N-1),$$

and so

$$\int_{\pi}^{\pi+\theta} \int_{z-2\theta}^{2\pi - z} \frac{\sin\left(\left(\frac{2N-1}{2}\right) z\right) \sin\left(\left(\frac{2N-1}{2}\right) w\right)}{\sin\left(\frac{z}{2}\right) \sin\left(\frac{w}{2}\right)} dwdz$$
$$\leq \frac{\pi^2}{\sqrt{2}} \left[\frac{1}{2} + \log(2\pi) + \log(2N-1)\right].$$

□

Having suffered through the analysis above, we find ourselves in a strong position: the eigenvalue counting function $\mathcal{N}_\theta$ of a random matrix from one of the classical compact groups is the sum of independent Bernoulli random variables, with explicit estimates on the mean and variance. We can thus bring all the results of classical probability to bear, starting with the central limit theorem.

Theorem 4.9 *Let $\mathcal{N}_\theta$ denote the eigenvalue counting function for a random matrix distributed in one of $\mathbb{U}(N)$, $\mathbb{SO}(2N)$, $\mathbb{SO}^-(2N+2)$, $\mathbb{SO}(2N+1)$, $\mathbb{SO}^-(2N+1)$, or $\mathbb{S}\mathbb{p}(2N)$, where $\theta \in [0, 2\pi)$ for $\mathbb{U}(N)$ and $\theta \in [0, \pi]$ otherwise. Then*

$$\lim_{N\to\infty} \mathbb{P}\left[\frac{\mathcal{N}_\theta - \mathbb{E}\mathcal{N}_\theta}{\operatorname{Var}(\mathcal{N}_\theta)} \leq t\right] = \frac{1}{\sqrt{2\pi}} \int_{-\infty}^{t} e^{-x^2/2} dx.$$

Proof Recall the Lindeberg central limit theorem: if for each n, $\{\xi_i\}_{i=1}^n$ are independent centered random variables with $s_n^2 = \operatorname{Var}\left(\sum_{i=1}^n \xi_i\right)$, then $\sum_{i=1}^n \xi_i$ satisfies a central limit theorem if for each $\epsilon > 0$,

$$\lim_{n\to\infty} \frac{1}{s_n^2} \sum_{i=1}^{n} \mathbb{E}\left[\xi_i^2 \mathbb{1}_{|\xi_i| \geq \epsilon s_n}\right] = 0. \tag{4.4}$$

While we did not state it explicitly, one can extract a lower bound on s_n^2 from the proof of Proposition 4.8, which shows that $\log(n)$ is the correct order. If $\mathcal{N}_\theta \stackrel{d}{=} \sum_{j=1}^{N} \eta_j$ with the η_j independent Bernoulli random variables as in Theorem 4.1, then taking $\xi_j = \eta_j - \mathbb{E}\eta_j$, condition (4.4) is satisfied: since $s_n \sim \sqrt{\log(n)}$ and the ξ_i are bounded, the expectations inside the sum are all zero for n large enough. □

Another classical result that is perhaps less familiar but will play an important role in later sections is the following.

Theorem 4.10 (Bernstein's inequality) *Let $\{X_j\}_{j=1}^n$ be independent random variables such that, for each j, $|X_j| \leq 1$ almost surely. Let $S_n := \sum_{j=1}^n X_j$ and let $\sigma^2 = \mathrm{Var}\left(\sum_{j=1}^n X_j\right)$. Then for all $t > 0$,*

$$\mathbb{P}\left[\left|S_n - \mathbb{E}[S_n]\right| > t\right] \leq 2\exp\left(-\frac{3t^2}{6\sigma^2 + 2t}\right).$$

Applying Bernstein's inequality to the eigenvalue counting functions and using the estimates for the mean and variance obtained above gives the following.

Theorem 4.11 *Let $\mathcal{N}_\theta$ denote the eigenvalue counting function for a random matrix distributed in one of $\mathbb{U}(N)$, $\mathbb{SO}(2N)$, $\mathbb{SO}^-(2N+2)$, $\mathbb{SO}(2N+1)$, $\mathbb{SO}^-(2N+1)$, or $\mathbb{S}\mathbb{p}(2N)$, where $\theta \in [0, 2\pi)$ for $\mathbb{U}(N)$ and $\theta \in [0, \pi]$ otherwise. Then there are constants C, c such that for all $t > 0$,*

$$\mathbb{P}\left[\left|\mathcal{N}_\theta - \frac{N\theta}{2\pi}\right| > t\right] \leq C\exp\left(-\frac{ct^2}{\log(N) + t}\right)$$

in the unitary case, and

$$\mathbb{P}\left[\left|\mathcal{N}_\theta - \frac{N\theta}{\pi}\right| > t\right] \leq C\exp\left(-\frac{ct^2}{\log(N) + t}\right)$$

in the remaining cases.

Consider, for example, $t = A\log(N)$. Writing $\epsilon_G = 2$ when the random matrix is coming from $G = \mathbb{U}(N)$ and $\epsilon_G = 1$ otherwise, Theorem 4.11 gives that

$$\mathbb{P}\left[\left|\mathcal{N}_\theta - \frac{N\theta}{\epsilon_G \pi}\right| > A\log(N)\right] \leq CN^{-\frac{cA^2}{A+1}};$$

that is, the $\mathcal{N}_\theta$ is extremely likely to be within a window of size $\log(N)$ around its mean, which is itself of order N.

4.2 The Empirical Spectral Measure and Linear Eigenvalue Statistics

This section introduces two of the main approaches to studying the asymptotic behavior of the eigenvalues: limit theorems for the empirical spectral measure and for linear eigenvalue statistics. We begin with the first of these two.

Definition Let U be a random matrix in one of the classical compact groups with eigenvalues $\lambda_1, \ldots, \lambda_n$. The **empirical spectral measure** μ_U is the random probability measure that puts equal mass at each of the eigenvalues of U:

$$\mu_U := \frac{1}{n} \sum_{j=1}^{n} \delta_{\lambda_j}.$$

Results on the asymptotic distribution of the eigenvalues of random matrices, as the size tends to infinity, are generally formulated in terms of the limiting behavior of the empirical spectral measure. Since this is a random measure, there are various possibilities. The weakest is **convergence in expectation**: A sequence μ_n of random probability measures on a metric measure space X converges in expectation to a deterministic limit μ if for any bounded, continuous $f : X \to \mathbb{R}$,

$$\lim_{n\to\infty} \mathbb{E}\left[\int f d\mu_n\right] = \int f d\mu. \tag{4.5}$$

For measures on $\mathbb{R}$, this is equivalent to the condition that

$$\lim_{n\to\infty} \mathbb{E}\mu_n((a,b]) = \mu((a,b]), \tag{4.6}$$

for all $a \le b$ such that $\mu(\{a\}) = \mu(\{b\}) = 0$.

A stronger notion is that of convergence **weakly in probability**: a sequence of random probability measures μ_n on $\mathbb{R}^n$ converge weakly in probability to a measure μ on $\mathbb{R}^n$ (written $\mu_n \xRightarrow{\mathbb{P}} \mu$) if for each bounded, continuous $f : \mathbb{R}^n \to \mathbb{R}$, the sequence of random variables $\int f d\mu_n$ converges in probability to $\int f d\mu$, as n tends to infinity.

There are several equivalent viewpoints; since our interest is in measures supported on the circle, we restrict our attention there for the following lemma.

Lemma 4.12 *For $j \in \mathbb{Z}$ and μ a probability measure on $[0, 2\pi)$, let $\hat{\mu}(j) = \int_0^{2\pi} e^{ij\theta} d\mu(\theta)$ denote the Fourier transform of μ at j. The following are equivalent:*

1. $\mu_n \xRightarrow{\mathbb{P}} \mu$;
2. *for each* $0 \le a \le b < 2\pi$ *such that* $\mu(\{a\}) = \mu(\{b\}) = 0$, $\mu_n((a,b]) \xrightarrow{\mathbb{P}} \mu((a,b])$;

3. *for each* $j \in \mathbb{Z}$, $\hat{\mu}_n(j) \xrightarrow[n\to\infty]{\mathbb{P}} \hat{\mu}(j)$;
4. *for every subsequence* n' *in* $\mathbb{N}$ *there is a further subsequence* n'' *such that with probability one,* $\mu_{n''} \Longrightarrow \mu$ *as* $n \to \infty$.

A still stronger notion of convergence is that of convergence **weakly almost surely**: a sequence of random probability measures μ_n on $\mathbb{R}^n$ converge weakly almost surely to a deterministic measure μ if the set on which μ_n converges weakly to μ has probability one.

We have seen that there are various metrics on the set of probability measures: the Kolmogorov distance, the bounded-Lipschitz distance, the L_p Kantorovich distances, the total variation distance. In addition to the types of convergence described above, one may of course consider the sequence of random variables $d(\mu_n, \mu)$ for any of these notions of distance, and show that it tends to zero (either weakly or in probability – when the limit is a point mass, they are equivalent).

Returning to the eigenvalue distribution of a random matrix, we saw in Proposition 4.7 that if U is a random unitary matrix, then for any $0 \le \theta_1 \le \theta_2 < 2\pi$,

$$\frac{1}{n}\mathbb{E}\mathcal{N}_{[\theta_1,\theta_2]} = \frac{\theta_2 - \theta_1}{2\pi},$$

and if U is a random matrix from any of the other groups, then for any $0 \le \theta_1 \le \theta_2 < \pi$,

$$\frac{1}{n}\mathbb{E}\mathcal{N}_{[\theta_1,\theta_2]} = \frac{\theta_2 - \theta_1}{\pi} + O\left(\frac{1}{n}\right).$$

That is, the empirical spectral measures all converge in expectation to the uniform measure on the circle.

In fact, the convergence happens in a much stronger sense.

Theorem 4.13 *Let* $\{\mu_n\}$ *be the empirical spectral measures of a sequence of random matrices* $\{U_n\}$, *with* U_n *drawn according to Haar measure on* $\mathbb{O}(n)$, $\mathbb{U}(n)$, *or* $\mathbb{S}\mathbb{p}(2n)$. *Let* ν *denote the uniform probability measure on the circle. Then with probability one,* μ_n *converges weakly to* ν *as* $n \to \infty$.

We will see a more recent approach to this result in Chapter 5; here, we present the original proof of Theorem 4.13 from [36], via the explicit moment formulae of Proposition 3.11. We will give the proof in the unitary case only; the others are analogous.

Proof of Theorem 4.13 Let $U \in \mathbb{U}(n)$ be a random unitary matrix with eigenvalues $\lambda_1, \ldots, \lambda_n$, and let $\mu_n := \frac{1}{n}\sum_{j=1}^n \delta_{\lambda_j}$ be the spectral measure of U. Let ν denote the uniform probability measure on the circle. We will show that μ_n

converges weakly to ν with probability one by showing that, with probability one, $\hat{\mu}_n(j) \to \hat{\nu}(j)$ for all $j \in \mathbb{Z}$. Now,

$$\hat{\mu}_n(j) = \int_{\mathbb{S}^1} z^j d\mu_n(z) = \frac{1}{n}\sum_{k=1}^{n} \lambda_k^j.$$

Since μ_n is a probability measure, $\hat{\mu}_n(0) = 1 = \hat{\nu}(0)$. Moreover, since λ^{-1} are the eigenvalues of $U^{-1} = U^*$, which is again Haar-distributed on $\mathbb{U}(n)$, it suffices to treat the case of $j \geq 1$. Now, Proposition 3.11 shows that

$$\mathbb{E}\hat{\mu}_n(j) = 0$$

and, using the fact that we may assume $n \geq j$, since j is fixed and $n \to \infty$,

$$\mathbb{E}|\hat{\mu}_n(j)|^2 = \frac{1}{n^2}\mathbb{E}\left[\mathrm{Tr}(U^j)\overline{\mathrm{Tr}(U^j)}\right] = \frac{j}{n^2}.$$

It thus follows from Chebychev's inequality that

$$\mathbb{P}[|\hat{\mu}_n(j)| > \epsilon_n] \leq \frac{j}{n^2\epsilon_n^2}$$

for any $\epsilon_n > 0$. In particular, taking $\epsilon_n = \frac{1}{n^{\frac{1}{2}-\delta}}$ for some $\delta \in \left(0, \frac{1}{2}\right)$ gives that

$$\mathbb{P}\left[|\hat{\mu}_n(j)| > \frac{1}{n^{\frac{1}{2}-\delta}}\right] \leq \frac{j}{n^{1+2\delta}},$$

which is summable, and so by the first Borel–Cantelli lemma, it holds with probability one that $|\hat{\mu}_n(j)| \leq \frac{1}{n^{\frac{1}{2}-\delta}}$ for n large enough. Taking the intersection of these probability one sets over all $j \in \mathbb{Z}$, we have that with probability one, for all $j \in \mathbb{Z}$, $\hat{\mu}_n(j) \to 0$ as $n \to \infty$. □

In addition to the proof given above of the convergence of the spectral measures, Proposition 3.11 is also a key tool in the study of linear eigenvalue statistics. By a **linear eigenvalue statistic**, we mean a function of U of the form

$$U \mapsto \sum_{j=1}^{n} f(\lambda_j),$$

where f is a test function and $\lambda_1, \ldots, \lambda_n$ are the eigenvalues of U. The class of test functions that works most naturally with the proof given below is the subspace $H_2^{\frac{1}{2}}$ of $L^2(\mathbb{S}^1)$ of those functions $f \in L^2(\mathbb{S}^1)$ with

$$\|f\|_{\frac{1}{2}}^2 = \sum_{j\in\mathbb{Z}} |\hat{f}_j|^2 |j| < \infty,$$

where $\hat{f}_j = \int_{\mathbb{S}^1} f(z)z^{-j}d\nu(z)$ is the jth Fourier coefficient of f. This space is an inner product space, with inner product given by

$$\langle f, g\rangle_{\frac{1}{2}} = \sum_{j\in\mathbb{Z}} \hat{f}_j\overline{\hat{g}_j}|j|.$$

For such test functions, there is the following multivariate central limit theorem for the corresponding linear eigenvalue statistics. To simplify the notation, given a test function f, let $\sigma_n(f) := \sum_{j=1}^n f(\lambda_j)$, where $\lambda_1, \dots, \lambda_n$ are the eigenvalues of a random unitary matrix $U \in \mathbb{U}(n)$.

Theorem 4.14 (Diaconis–Evans [32]) *Let $f_1, \dots, f_k \in H_2^{\frac{1}{2}}$, and suppose that $\mathbb{E}[\sigma_n(f_j)] = 0$ for each $j \in 1, \dots, k$. The random vector $(\sigma_n(f_1), \dots, \sigma_n(f_k))$ converges in distribution to a jointly Gaussian random vector $(Y_1, \dots, Y_k)$, with $\mathbb{E}[Y_i] = 0$ for each i, and $\mathbb{E}[Y_iY_j] = \langle f_i, f_j\rangle_{\frac{1}{2}}$.*

The proof uses the following multivariate central limit theorem for the traces of powers of U. It is an immediate consequence of Proposition 3.11 that for fixed k, the random vector $(\mathrm{Tr}(U), \dots, \mathrm{Tr}(U^k))$ converges in distribution to $(Z_1, \sqrt{2}Z_2, \dots, \sqrt{k}Z_k)$, where $Z_1, \dots, Z_k$ are independent standard complex Gaussian random variables, but to prove Theorem 4.14, the following stronger result is needed.

Theorem 4.15 *Let $\{a_{nj}\}_{n,j\in\mathbb{N}}$ and $\{b_{nj}\}_{n,j\in\mathbb{N}}$ be arrays of complex constants. Suppose that there exist σ^2, τ^2, and γ such that*

$$\lim_{n\to\infty}\sum_{j=1}^{\infty}|a_{nj}|^2\min\{j,n\} = \sigma^2 \qquad \lim_{n\to\infty}\sum_{j=1}^{\infty}|b_{nj}|^2\min\{j,n\} = \tau^2$$

$$\lim_{n\to\infty}\sum_{j=1}^{\infty}a_{nj}b_{nj}\min\{j,n\} = \gamma.$$

Suppose moreover that there is a sequence $(m_n)_{n\geq 1} \subseteq \mathbb{N}$ such that $\lim_{n\to\infty}\frac{m_n}{n} = 0$ and such that

$$\lim_{n\to\infty}\sum_{j=m_n+1}^{\infty}(|a_{nj}|^2 + |b_{nj}|^2)\min\{j,n\} = 0.$$

Then the random variable $\sum_{j=1}^{\infty}(a_{nj}\,\mathrm{Tr}(U^j) + b_{nj}\overline{\mathrm{Tr}(U^j)})$ converges in distribution (as $n\to\infty$) to a centered complex Gaussian random variable $X + iY$, with

$$\mathbb{E}X^2 = \frac{1}{2}(\sigma^2+\tau^2+2\,\mathrm{Re}(\gamma)) \qquad \mathbb{E}Y^2 = \frac{1}{2}(\sigma^2+\tau^2-2\,\mathrm{Re}(\gamma))$$

$$\mathbb{E}XY = \mathrm{Im}(\gamma).$$

Proof From Theorem 3.11, $\mathbb{E}\,\mathrm{Tr}(U^j) = 0$ for each j. For $N < M$,

$$\mathbb{E}\left|\sum_{j=N}^{M}(a_{nj}\,\mathrm{Tr}(U^j) + b_{nj}\overline{\mathrm{Tr}(U^j)})\right|^2 = \sum_{j=N}^{M}(|a_{nj}|^2 + |b_{nj}|^2)\min\{j,n\} \xrightarrow{N,M\to\infty} 0,$$

so that $\sum_{j=1}^{\infty}(a_{nj}\,\mathrm{Tr}(U^j) + b_{nj}\overline{\mathrm{Tr}(U^j)})$ converges in L^2. Similarly,

$$\mathbb{E}\left|\sum_{j=m_n+1}^{\infty}(a_{nj}\,\mathrm{Tr}(U^j) + b_{nj}\overline{\mathrm{Tr}(U^j)})\right|^2 = \sum_{j=m_n+1}^{\infty}(|a_{nj}|^2+|b_{nj}|^2)\min\{j,n\} \xrightarrow{n\to\infty} 0,$$

so that $\sum_{j=m_n+1}^{\infty}(a_{nj}\,\mathrm{Tr}(U^j) + b_{nj}\overline{\mathrm{Tr}(U^j)}) \Longrightarrow 0$.

It is thus enough to show that $\sum_{j=1}^{m_n}(a_{nj}\,\mathrm{Tr}(U^j) + b_{nj}\overline{\mathrm{Tr}(U^j)}) \Longrightarrow X + iY$, for which we use the method of moments. Since $\frac{m_n}{n} \to 0$, if $\alpha, \beta \in \mathbb{N}$ are fixed, then $\alpha m_n, \beta m_n \leq n$ for n large enough, and for such α, β, it follows from Theorem 3.11 that

$$\mathbb{E}\left[\left(\sum_{j=1}^{m_n}(a_{nj}\,\mathrm{Tr}(U^j) + b_{nj}\overline{\mathrm{Tr}(U^j)})\right)^{\alpha}\overline{\left(\sum_{j=1}^{m_n}(a_{nj}\,\mathrm{Tr}(U^j) + b_{nj}\overline{\mathrm{Tr}(U^j)})\right)^{\beta}}\right]$$
$$= \mathbb{E}\left[\left(\sum_{j=1}^{m_n}(a_{nj}\sqrt{j}Z_j + b_{nj}\sqrt{j}\overline{Z_j})\right)^{\alpha}\overline{\left(\sum_{j=1}^{m_n}(a_{nj}\sqrt{j}Z_j + b_{nj}\sqrt{j}\overline{Z_j})\right)^{\beta}}\right].$$

Writing $Z_j = \frac{1}{\sqrt{2}}(Z_{j1} + iZ_{j2})$ with the Z_{ji} i.i.d. standard Gaussians,

$$\sum_{j=1}^{m_n}(a_{nj}\sqrt{j}Z_j + b_{nj}\sqrt{j}\overline{Z_j}) = X_n + iY_n$$

with

$$X_n = \sum_{j=1}^{m_n}\sqrt{\frac{j}{2}}\left[\left(\mathrm{Re}(a_{nj}) + \mathrm{Re}(b_{nj})\right)Z_{j1} + \left(\mathrm{Im}(b_{nj}) - \mathrm{Im}(a_{nj})\right)Z_{j2}\right]$$

and

$$Y_n = \sum_{j=1}^{m_n}\sqrt{\frac{j}{2}}\left[\left(\mathrm{Im}(a_{nj}) + \mathrm{Im}(b_{nj})\right)Z_{j1} + \left(\mathrm{Re}(a_{nj}) - \mathrm{Re}(b_{nj})\right)Z_{j2}\right].$$

The random variables X_n and Y_n are thus centered and jointly Gaussian, with

$$\begin{aligned}\mathbb{E}X_n^2 &= \sum_{j=1}^{m_n} \left(\frac{j}{2}\right) \left[(\mathrm{Re}(a_{nj}) + \mathrm{Re}(b_{nj}))^2 + (\mathrm{Im}(a_{nj}) - \mathrm{Im}(b_{nj}))^2 \right] \\ &= \sum_{j=1}^{m_n} \left(\frac{j}{2}\right) \left[|a_{nj}|^2 + |b_{nj}|^2 + 2\,\mathrm{Re}(a_{nj}b_{nj}) \right], \\ \mathbb{E}Y_n^2 &= \sum_{j=1}^{m_n} \left(\frac{j}{2}\right) \left[(\mathrm{Re}(a_{nj}) - \mathrm{Re}(b_{nj}))^2 + (\mathrm{Im}(a_{nj}) + \mathrm{Im}(b_{nj}))^2 \right] \\ &= \sum_{j=1}^{m_n} \left(\frac{j}{2}\right) \left[|a_{nj}|^2 + |b_{nj}|^2 - 2\,\mathrm{Re}(a_{nj}b_{nj}) \right],\end{aligned}$$

and

$$\begin{aligned}\mathbb{E}X_nY_n = \sum_{j=1}^{m_n} \left(\frac{j}{2}\right) \Big[&(\mathrm{Re}(a_{nj}) + \mathrm{Re}(b_{nj}))(\mathrm{Im}(a_{nj}) + \mathrm{Im}(b_{nj})) \\ &- (\mathrm{Re}(a_{nj}) - \mathrm{Re}(b_{nj}))(\mathrm{Im}(a_{nj}) - \mathrm{Im}(b_{nj})) \Big] \\ = \sum_{j=1}^{m_n} &j\,\mathrm{Im}(a_{nj}b_{nj}).\end{aligned}$$

It thus follows by the assumptions on the arrays $\{a_{nj}\}$ and $\{b_{nj}\}$ and the sequence (m_n) that

$$\mathbb{E}X_n \to \frac{1}{2}(\sigma^2 + \tau^2 + 2\,\mathrm{Re}(\gamma)) \qquad \mathbb{E}Y_n \to \frac{1}{2}(\sigma^2 + \tau^2 - 2\,\mathrm{Re}(\gamma))$$

and

$$\mathbb{E}X_nY_n \to \mathrm{Im}(\gamma),$$

which completes the proof. □

Proof of Theorem 4.14 By the Cramér–Wold device, to prove the claimed convergence it is enough to show that

$$\sum_{j=1}^{k} t_j\sigma_n(f_j) \xrightarrow{d} \sum_{j=1}^{k} t_jY_j$$

for each $(t_1, \dots, t_k) \in \mathbb{R}^k$. Note that

$$\sum_{j=1}^{k} t_j\sigma_n(f_j) = \sigma_n(F_t),$$

where $F_t = \sum_{j=1}^{k} t_j f_j$. Since $H_2^{\frac{1}{2}}$ is a vector space, $F_t \in H_2^{\frac{1}{2}}$, and moreover,

$$\|F_t\|_{\frac{1}{2}} = \sum_{j,\ell=1}^{k} t_j t_\ell \left\langle f_j, f_\ell \right\rangle_{\frac{1}{2}} = \mathrm{Var}\left(\sum_{j=1}^{k} Y_j\right).$$

The theorem therefore follows from the $k = 1$ case.

The $k = 1$ case is simple if $f(z) = \sum_{j=-N}^{N} a_j z^j$ is a Laurent polynomial on $\mathbb{S}^1$. Since we are assuming that f is real-valued and that $\sigma_n(f) = 0$, we have that $a_0 = 0$ and $a_{-j} = \overline{a_j}$. Then

$$\sigma_n(f) = 2\sum_{j=1}^{N}\left[\mathrm{Re}(a_j)\,\mathrm{Re}(\mathrm{Tr}(U^j)) - \mathrm{Im}(a_j)\,\mathrm{Im}(\mathrm{Tr}(U^j))\right],$$

which tends to the Gaussian random variable

$$2\sum_{j=1}^{N}\left[\mathrm{Re}(a_j)\sqrt{\frac{j}{2}}Z_{j1} - \mathrm{Im}(a_j)\sqrt{\frac{j}{2}}Z_{j2}\right],$$

where $(Z_{11}, Z_{12}, \dots, Z_{k1}, Z_{k2})$ are i.i.d. standard (real) Gaussians. That is, $\sigma_n(f)$ converges to a centered Gaussian random variable with variance

$$\sigma^2 = 4\sum_{j=1}^{N}\left(\frac{j}{2}\right)(\mathrm{Re}(a_j)^2 + \mathrm{Im}(a_j)^2) = \sum_{j=-N}^{N} |j||a_j|^2 = \|f\|_{\frac{1}{2}}^2.$$

For a general $f \in H_2^{\frac{1}{2}}$, the full strength of Theorem 4.15 is needed. First, for $N \in \mathbb{N}$ define

$$f_N := \sum_{j=-N}^{N} \hat{f}_j z^j$$

and note that, since $f \in L^2(\mathbb{S}^1)$, $\|f_N - f\|_2 \to 0$. It follows from symmetry that for any measurable subset $A \subseteq \mathbb{S}^1$, the probability that U has at least one eigenvalue in A is at most $n\nu(A)$, where ν is the uniform probability measure on $\mathbb{S}^1$, and so if $A_{n,\epsilon}$ denotes the set on which $|f - f_N| > \frac{\epsilon}{n}$, then

$$\begin{aligned}\mathbb{P}[|\sigma_n(f) - \sigma_n(f_N)| > \epsilon] = \mathbb{P}\left[\left|\sum_{\ell=1}^{n}(f(\lambda_\ell) - f_n(\lambda_\ell))\right| > \epsilon\right] &\leq n\nu(A_{n,\epsilon}) \\ &\leq \frac{n^3\|f - f_N\|_2^2}{\epsilon^2},\end{aligned}$$

which tends to zero as N tends to infinity. That is, $\sigma_n(f_N)$ converges in probability to $\sigma_n(f)$, as N tends to infinity.

On the other hand, the condition on f together with Theorem 3.11 means that

$$\sum_{j=1}^{\infty} \hat{f}_j \operatorname{Tr}(U^j) + \sum_{j=1}^{\infty} \overline{\hat{f}_j \operatorname{Tr}(U^j)}$$

converges in L^2:

$$\mathbb{E}\left|\sum_{j=N}^{M} \hat{f}_j \operatorname{Tr}(U^j)\right|^2 = \sum_{j=N}^{M} |\hat{f}_j|^2 \min\{j,n\} \xrightarrow{N,M\to\infty} 0.$$

This allows us to write

$$\sigma_n(f) = \sum_{j=1}^{\infty} \hat{f}_j \operatorname{Tr}(U^j) + \sum_{j=1}^{\infty} \overline{\hat{f}_j \operatorname{Tr}(U^j)}.$$

(Note that the corresponding formula for $\sigma_n(f)$ in terms of traces of powers of U was trivial in the case that f was a Laurent polynomial because the sums needing to be interchanged were both finite, so that convergence issues played no role.)

We now apply Theorem 4.15 with $a_{jn} = \hat{f}_j$ and $b_{jn} = \overline{\hat{f}_j}$:

$$\begin{aligned}\sum_{j=1}^{\infty} |a_{jn}|^2 \min\{j,n\} = \sum_{j=1}^{\infty} |b_{jn}|^2 \min\{j,n\} &= \sum_{j=1}^{\infty} a_{nj} b_{nj} \min\{j,n\} \\ &= \sum_{j=1}^{\infty} |\hat{f}_j|^2 j + \sum_{j=n+1}^{\infty} (n-j)|\hat{f}_j|^2 \xrightarrow{n\to\infty} \|f\|_{\frac{1}{2}}^2,\end{aligned}$$

and as long as $m_n \to \infty$, then

$$\sum_{j=m_n+1}^{\infty} (|a_{nj}|^2 + |b_{nj}|^2) \min\{j,n\} = \sum_{j=m_n+1}^{\infty} (2|\hat{f}_j|^2) \min\{j,n\} \xrightarrow{n\to\infty} 0.$$

It thus follows from Theorem 4.15 that $\sigma_n(f)$ converges to a centered Gaussian random variable with variance $\|f\|_{\frac{1}{2}}^2$, as desired. □

4.3 Patterns in Eigenvalues: Self-Similarity

Consider the special case of $m = 2$ in Theorem 3.14: the eigenvalues of the square of a $2n \times 2n$ random unitary matrix are identical in distribution to those of two independent $n \times n$ random unitary matrices. Since the squaring function has the effect of wrapping the circle in the complex plane twice around, this means that wrapping the eigenvalues of a $2n \times 2n$ unitary matrix twice around the circle produces two independent copies of the eigenvalues of a half-sized random matrix. It is tempting to see those two copies as corresponding to the eigenvalues of the original matrix from the top half of the circle and those from the bottom half; if this were true, it would be a kind of self-similarity statement for unitary eigenvalues. Of course, there cannot be exact equality of distributions due to edge effects, but one could hope for some weaker result along these lines which avoided such effects. Indeed, a closely related self-similarity phenomenon was conjectured by Coram and Diaconis in their statistical analysis of unitary eigenvalues [24]: they asked whether taking half (sequentially) of the eigenvalues and stretching them out to fill the whole circle would produce something "statistically indistinguishable" from the eigenvalues of a half-sized matrix.

The following result gives exactly such a self-similarity phenomenon on a certain mesoscopic scale.

Theorem 4.16 *For $m, n \geq 1$, let $\{\theta_j\}_{j=1}^n$ be the eigenangles of a random $n \times n$ unitary matrix and let $\{\theta_j^{(m)}\}_{j=1}^{nm}$ be the eigenangles of a random $nm \times nm$ unitary matrix, with $\theta_j, \theta_j^{(m)} \in [-\pi, \pi)$ for each j. For $A \subseteq [-\pi, \pi)$, let*

$$\mathcal{N}_A := \left|\{j : \theta_j \in A\}\right| \qquad \mathcal{N}_A^{(m)} := \left|\left\{j : \theta_j^{(m)} \in \left[-\frac{\pi}{m}, \frac{\pi}{m}\right), m\theta_j^{(m)} \in A\right\}\right|;$$

that is, $\mathcal{N}_A$ is the counting function of a random $n \times n$ unitary matrix, and $\mathcal{N}_A^{(m)}$ is the counting function for the point process obtained by taking the eigenvalues of a random $nm \times nm$ unitary matrix lying in the arc of length $\frac{2\pi}{m}$ centered at $z = 1$ and raising them to the mth power.

Suppose that A has $\operatorname{diam}(A) \leq \pi$. *Then*

$$d_{TV}(\mathcal{N}_A, \mathcal{N}_A^{(m)}) \leq W_1(\mathcal{N}_A, \mathcal{N}_A^{(m)}) \leq \frac{\sqrt{mn}|A| \operatorname{diam}(A)}{12\pi},$$

where $|A|$ denotes the Lebesgue measure of A.

As a consequence of Theorem 4.16, if $\{A_n\}$ is a sequence of sets such that either $\operatorname{diam} A_n = o(n^{-1/4})$ or $|A_n| = o(n^{-1/2})$ as $n \to \infty$, then

$$d_{TV}\left(\mathcal{N}_{A_n}, \mathcal{N}_{A_n}^{(m)}\right), W_1\left(\mathcal{N}_{A_n}, \mathcal{N}_{A_n}^{(m)}\right) \to 0.$$

Thus indeed, a sequential arc of about n of the nm eigenvalues of an $nm \times nm$ random matrix is statistically indistinguishable, on the scale of $o(n^{-1/4})$ for diameter or $o(n^{-1/2})$ for Lebesgue measure, from the n eigenvalues of an $n \times n$ random matrix.

How sharp the bound above is remains an open question. It seems likely that the restriction on the diameter of A is an artifact of the proof; this may also be the case for the factor of $\sqrt{n}$ in the bound.

Nevertheless, a remarkable feature of Theorem 4.16 is that it yields *microscopic* information even at a *mesoscopic* scale: if $\frac{1}{n} \ll \operatorname{diam} A_n \ll \frac{1}{n^{1/4}}$, then $\mathcal{N}_{A_n}$ and $\mathcal{N}_{A_n}^{(m)}$ both have expectations and variances tending to infinity. Typically, there is no hope in looking at individual points in such a setting, and instead one studies statistical properties of the recentered and rescaled counts. Theorem 4.16 makes a direct point-by-point comparison of the two point processes, with no rescaling.

The proof of Theorem 4.16 is based on the determinantal structure of the two point processes. Let χ denote the point process given by the eigenvalues of an $n \times n$ random unitary matrix, and let $\chi^{(m)}$ be the process given by restricting the eigenvalue process of a random $mn \times mn$ matrix to $\left[\frac{-\pi}{m}, \frac{\pi}{m}\right)$ and rescaling by m. Recall from Section 3.1 that χ is a determinantal point process; it follows easily that $\chi^{(m)}$ is as well, with kernel as follows.

Proposition 4.17 *The point process $\chi^{(m)}$ on $[0, 2\pi)$ is determinantal with kernel*

$$K_n^{(m)}(x, y) = \frac{1}{2\pi} \frac{\sin\left(\frac{n(x-y)}{2}\right)}{m \sin\left(\frac{(x-y)}{2m}\right)}.$$

with respect to Lebesgue measure.

Proof The case $m = 1$ is given in Section 3.1. The general case follows from a change of variables which shows that $K_n^{(m)}(x, y) = K_{mn}\left(\frac{x}{m}, \frac{y}{m}\right)$. □

The main technical ingredient for the proof of Theorem 4.16 is the following general result on determinantal point processes.

Proposition 4.18 *Let $\mathcal{N}$ and $\widetilde{\mathcal{N}}$ be the total numbers of points in two determinantal point processes on (Λ, μ) with conjugate-symmetric kernels $K, \widetilde{K} \in L^2(\mu \otimes \mu)$, respectively. Suppose that $\mathcal{N}, \widetilde{\mathcal{N}} \leq N$ almost surely. Then*

$$d_{TV}(\mathcal{N}, \widetilde{\mathcal{N}}) \leq W_1(\mathcal{N}, \widetilde{\mathcal{N}}) \leq \sqrt{N \int\int \left|K(x, y) - \widetilde{K}(x, y)\right|^2 d\mu(x) d\mu(y)}.$$

Proof Note first that an indicator function of a set A of integers is 1-Lipschitz on $\mathbb{Z}$, and so for X and Y integer-valued,

$$d_{TV}(X, Y) \leq W_1(X, Y); \tag{4.7}$$

it therefore suffices to prove the second inequality.

Let $\{\lambda_j\}$ and $\{\widetilde{\lambda}_j\}$ be the eigenvalues, listed in nonincreasing order, of the integral operators $\mathcal{K}$ and $\widetilde{\mathcal{K}}$ with kernels K and $\widetilde{K}$, respectively. Since $\mathcal{N}, \widetilde{\mathcal{N}} \leq N$, by Theorem 4.1 we may assume that $j \leq N$. Let $\{Y_j\}_{j=1}^N$ be independent random variables uniformly distributed in $[0, 1]$. For each j, define

$$\xi_j = \mathbb{1}_{Y_j \leq \lambda_j} \qquad \text{and} \qquad \widetilde{\xi}_j = \mathbb{1}_{Y_j \leq \widetilde{\lambda}_j}.$$

It follows from Theorem 4.1 that

$$\mathcal{N} \stackrel{d}{=} \sum_{j=1}^N \xi_j, \qquad \widetilde{\mathcal{N}} \stackrel{d}{=} \sum_{j=1}^N \widetilde{\xi}_j,$$

so this gives a coupling of $\mathcal{N}$ and $\widetilde{\mathcal{N}}$. It then follows from the Kantorovich–Rubenstein theorem that

$$\begin{aligned} W_1(\mathcal{N}, \widetilde{\mathcal{N}}) &\leq \mathbb{E} \left| \sum_{j=1}^N \xi_j - \sum_{j=1}^N \widetilde{\xi}_j \right| \leq \sum_{j=1}^N \mathbb{E} \left| \xi_j - \widetilde{\xi}_j \right| \\ &= \sum_{j=1}^N \left| \lambda_j - \widetilde{\lambda}_j \right| \leq \sqrt{N \sum_{j=1}^N \left| \lambda_j - \widetilde{\lambda}_j \right|^2}. \end{aligned} \tag{4.8}$$

For $n \times n$ Hermitian matrices A and B, the Hoffman–Wielandt inequality (see, e.g., [11, theorem VI.4.1]) says that

$$\sum_{j=1}^n \left| \lambda_j(A) - \lambda_j(B) \right|^2 \leq \|A - B\|_{HS}^2, \tag{4.9}$$

where $\lambda_1(A), \ldots, \lambda_n(A)$ and $\lambda_1(B), \ldots, \lambda_n(B)$ are the eigenvalues (in nonincreasing order) of A and B respectively. It thus follows that

$$\sqrt{\sum_{j=1}^N \left| \lambda_j - \widetilde{\lambda}_j \right|^2} \leq \left\| \mathcal{K} - \widetilde{\mathcal{K}} \right\|_{H.S.},$$

where $\|\cdot\|_{H.S.}$ denotes the Hilbert–Schmidt norm. The result now follows from the general fact that the Hilbert–Schmidt norm of an integral operator on $L^2(\mu)$ is given by the $L^2(\mu \otimes \mu)$ norm of its kernel (see [107, pg. 245]). □

Proof of Theorem 4.16 For every $0 \leq \varphi \leq \frac{\pi}{2}$,

$$\varphi - \frac{1}{6}\varphi^3 \leq \sin \varphi \leq m \sin\left(\frac{\varphi}{m}\right) \leq \varphi,$$

and so

$$0 \le \frac{1}{\sin\varphi} - \frac{1}{m\sin\left(\frac{\varphi}{m}\right)} \le \frac{1}{\varphi - \frac{1}{6}\varphi^3} - \frac{1}{\varphi} = \frac{\varphi}{6-\varphi^2} \le \frac{\varphi}{3}.$$

Thus by Propositions 4.17 and 4.18,

$$\begin{aligned} W_1\left(\mathcal{N}_A, \mathcal{N}_A^{(m)}\right) &\le \sqrt{\frac{mn}{(2\pi)^2}\int_A\int_A \sin^2\left(\frac{n(x-y)}{2}\right)\left(\frac{1}{\sin\left(\frac{x-y}{2}\right)} - \frac{1}{m\sin\left(\frac{x-y}{2m}\right)}\right)^2 dx\,dy} \\ &\le \frac{1}{12\pi}\sqrt{mn\int_A\int_A (x-y)^2\,dx\,dy} \\ &\le \frac{1}{12\pi}\sqrt{mn}\,|A|\operatorname{diam} A. \end{aligned}$$

□

4.4 Large Deviations for the Empirical Spectral Measure

Large deviations theory is a branch of probability dealing with rare events. The basic idea is the following. Suppose that $X_1, X_2, \ldots$ are i.i.d. random variables with $\mathbb{E}X_1 = 0$ and $\mathbb{E}X_1^2 = 1$, and let $S_n := \sum_{j=1}^n X_j$. By the law of large numbers, $\frac{1}{n}S_n$ should be close to zero, but of course it is not typically exactly zero; the goal is to understand the rare event that $\left|\frac{1}{n}S_n\right| > \delta$ for some $\delta > 0$ fixed. By the central limit theorem, $\frac{1}{n}S_n$ is approximately distributed as a centered Gaussian random variable with variance $\frac{1}{n}$, and so if Z denotes a standard Gaussian random variable, then

$$\mathbb{P}\left[\left|\frac{1}{n}S_n\right| > \delta\right] \approx \mathbb{P}[|Z| > \delta\sqrt{n}] = \frac{2}{\sqrt{2\pi}}\int_{\delta\sqrt{n}}^\infty e^{-\frac{x^2}{2}}\,dx.$$

Using standard asymptotics for the tail of a Gaussian integral,

$$\frac{1}{n}\log\mathbb{P}\left[|Z| > \delta\sqrt{n}\right] \xrightarrow{n\to\infty} -\frac{\delta^2}{2}.$$

This limit can be loosely interpreted as saying that the probability that $\left|\frac{1}{\sqrt{n}}Z\right| > \delta$ is of the order $e^{-\frac{n\delta^2}{2}}$.

The question is then whether it also holds that

$$\frac{1}{n}\log\mathbb{P}\left[\left|\frac{1}{n}S_n\right| > \delta\right] \xrightarrow{n\to\infty} -\frac{\delta^2}{2}. \tag{4.10}$$

The answer (given by Cramér's theorem) is "sort of": the limit in (4.10) exists, but rather than being given by a universal function of δ, its value depends on the distribution of the summands.

Statements like (4.10) are what is meant when one refers to *large deviations*, although typical theorems are formulated somewhat differently. Below, we give the basic definitions and terminology used in the context of large deviations for sequences of Borel probability measures on a topological space X.

Definitions 1. A **rate function** I is a lower semicontinuous[1] function $I : X \to [0, \infty]$.

2. A **good rate function** is a rate function for which all level sets $I^{-1}([0, \alpha])$ $(\alpha < \infty)$ are compact.
3. A sequence of Borel measures $\{\nu_n\}_{n\in\mathbb{N}}$ on X satisfies a **large deviations principle (LDP)** with rate function I and **speed** s_n if for all Borel sets $\Gamma \subseteq X$,

$$\begin{aligned} -\inf_{x\in\Gamma^\circ} I(x) &\le \liminf_{n\to\infty} \frac{1}{s_n} \log(\nu_n(\Gamma)) \\ &\le \limsup_{n\to\infty} \frac{1}{s_n} \log(\nu_n(\Gamma)) \le -\inf_{x\in\overline{\Gamma}} I(x), \end{aligned} \tag{4.11}$$

where Γ° is the interior of Γ and $\overline{\Gamma}$ is the closure of Γ.

If the upper bound in (4.11) is required to hold only when Γ is compact, the sequence $\{\nu_n\}_{n\in\mathbb{N}}$ satisfies a **weak LDP**.

The somewhat complicated form of a large deviations principle, as compared with, e.g., (4.10), has the advantage of being precise enough to be useful while weak enough to have some chance of holding in situations of interest.

The following gives an indirect approach for establishing the existence of a weak LDP.

Theorem 4.19 *Let $\{\nu_n\}_{n\in\mathbb{N}}$ be a sequence of Borel measures on X and, and for each n, let $s_n > 0$. Let $\mathcal{A}$ be a base for the topology of X. For $x \in X$, define*

$$I(x) := \sup_{A\in\mathcal{A}:x\in A} \left[-\liminf_{n\to\infty} \frac{1}{s_n} \log \nu_n(A) \right].$$

Suppose that for all $x \in X$,

$$I(x) = \sup_{A\in\mathcal{A}:x\in A} \left[-\limsup_{n\to\infty} \frac{1}{s_n} \log \nu_n(A) \right].$$

Then $\{\nu_n\}_{n\in\mathbb{N}}$ satisfies a weak LDP with rate function I and speed s_n.

[1] i.e., for all $\alpha \in [0, \infty)$, the level set $I^{-1}([0, \alpha])$ is closed in X.

The Logarithmic Energy

Let ν be a measure on $\mathbb{C}$. The quantity

$$\mathcal{E}(\nu) := -\iint \log|z-w|d\nu(z)d\nu(w)$$

is called (in potential theory) the **logarithmic energy** of ν; the same quantity appears in free probability as (the negative of) the **free entropy**. As suggested by the following lemma, this energy functional will play a key role in the LDP for empirical spectral measures.

Lemma 4.20 *Let $\mathcal{P}(\mathbb{S}^1)$ denote the space of Borel probability measures on $\mathbb{S}^1$, endowed with the topology of convergence in distribution. The logarithmic energy $\mathcal{E}$ is a strictly convex rate function on $\mathcal{P}(\mathbb{S}^1)$ and the uniform probability measure ν_0 is the unique $\nu \in \mathcal{P}(\mathbb{S}^1)$ with $\mathcal{E}(\nu) = 0$.*

Proof First, if ν_0 is the uniform probability measure on $\mathbb{S}^1$, then

$$\begin{aligned}\mathcal{E}(\nu_0) &= -\frac{1}{(2\pi)^2}\int_0^{2\pi}\int_0^{2\pi}\log|e^{i\theta}-e^{i\phi}|d\theta d\phi \\ &= -\frac{1}{2\pi}\int_0^{2\pi}\log|1-e^{i\theta}|d\theta = 0.\end{aligned}$$

The fact that ν_0 is the unique element $\nu \in \mathcal{P}(\mathbb{S}^1)$ with $\mathcal{E}(\nu) = 0$ will be shown below.

Both the nonnegativity and strict convexity of $\mathcal{E}$ are most easily seen within the context of positive- and negative-definite kernels. A Hermitian kernel $L(x, y)$ on $\mathbb{C} \times \mathbb{C}$ is called **positive definite** if

$$\sum_{j,k} c_j\overline{c_k}L(x_j, x_k) \geq 0$$

for all $c_1, \ldots, c_n \in \mathbb{C}$. A Hermitian kernel $L(x, y)$ on $\mathbb{C} \times \mathbb{C}$ is called **(conditionally) negative definite** if

$$\sum_{j,k} c_j\overline{c_k}L(x_j, x_k) \leq 0$$

for all $c_1, \ldots, c_n \in \mathbb{C}$ such that $\sum_{j=1}^n c_j = 0$. Obviously the negative of a positive definite kernel is negative definite, but the negative of a negative definite kernel need not be positive definite. Note that sums or integrals of positive (resp. negative) definite kernels are positive (resp. negative) definite.

Given a negative-definite kernel on $\mathbb{C} \times \mathbb{C}$ and a finite signed measure μ on $\mathbb{C}$ with $\mu(\mathbb{C}) = 0$, it follows by approximating μ by discrete measures that

$$\iint L(z, w) d\mu(x) d\mu(w) \leq 0.$$

Our interest is in this last statement, for the singular kernel $K(z, w) = \log |z-w|$. To avoid the singularity, define for each $\epsilon > 0$ the kernel

$$K_\epsilon(z, w) := \log(\epsilon + |z - w|) = \int_0^\infty \left(\frac{1}{1+t} - \frac{1}{t + \epsilon + |z - w|} \right) dt. \quad (4.12)$$

Now, the kernel

$$(z, w) \mapsto \frac{1}{t + \epsilon + |z - w|}$$

is positive definite for each t and ϵ: the Laplacian kernel $(z, w) \mapsto e^{-s|z-w|}$ is known to be positive definite for each $s > 0$, and

$$\frac{1}{t + \epsilon + |z - w|} = \int_0^\infty e^{-s(t+\epsilon+|z-w|)} ds$$

is therefore the integral of positive definite kernels, hence positive definite. It is easy to see that a constant kernel is conditionally negative definite, and so the integrand in (4.12) is conditionally negative definite for each t, which finally gives that $K_\epsilon(z, w)$ is conditionally negative definite for each $\epsilon > 0$. It thus follows that

$$\iint K_\epsilon(z, w) d\mu(z) d\mu(w) \leq 0.$$

For $\epsilon < \frac{1}{2}$, $|K_\epsilon(z, w)| \leq |K(z, w)| + \log(2)$, and so if $\mathcal{E}(\mu) < \infty$, then it follows by the dominated convergence theorem that

$$\mathcal{E}(\mu) = -\iint K(z, w) d\mu(z) d\mu(w) \geq 0.$$

We are, of course, not interested in signed measures of total mass 0, but in probability measures on $\mathbb{S}^1$. Given a probability measure ν on $\mathbb{S}^1$, let $\mu = \nu - \nu_0$, where ν_0 is the uniform probability measure on $\mathbb{S}^1$. Then by the argument above,

$$\iint K(z, w) d\mu(z) d\mu(w) = -\mathcal{E}(\nu) - 2 \iint K(z, w) d\nu(z) d\nu_0(w) - \mathcal{E}(\nu_0) \leq 0.$$

It has already been shown that $\mathcal{E}(\nu_0) = 0$, and so the above inequality reduces to

$$\mathcal{E}(\nu) \geq -2 \iint K(z,w)d\nu(z)d\nu_0(w).$$

But

$$\iint K(z,w)d\nu(z)d\nu_0(w) = \int_{\mathbb{S}^1} \left(\frac{1}{2\pi} \int_0^{2\pi} \log|z - e^{i\theta}|d\theta \right) d\nu(z) = 0,$$

which proves the nonnegativity of $\mathcal{E}(\nu)$.

To prove the convexity of $\mathcal{E}$, let ν_1 and ν_2 be distinct probability measures on $\mathbb{S}^1$ with finite logarithmic energy. By a similar argument to the one proving the nonnegativity of $\mathcal{E}$,

$$\begin{aligned}\iint & K(z,w)d(\nu_1 - \nu_2)(z)d(\nu_1 - \nu_2)(w) \\ &= -\mathcal{E}(\nu_1) - 2 \iint K(z,w)d\nu_1(z)d\nu_2(w) - \mathcal{E}(\nu_2) \leq 0,\end{aligned}$$

so that

$$\mathcal{E}(\nu_1, \nu_2) := -\iint \log|z - w|d\nu_1(z)d\nu_2(w) \leq \frac{1}{2}(\mathcal{E}(\nu_1) + \mathcal{E}(\nu_2)) < \infty.$$

Moreover, for $0 < \lambda < 1$,

$$\mathcal{E}(\lambda\nu_1 + (1-\lambda)\nu_2) = \mathcal{E}(\nu_2) + 2\lambda\mathcal{E}(\nu_2, \nu_1 - \nu_2) + \lambda^2\mathcal{E}(\nu_1 - \nu_2).$$

It follows that

$$\frac{d^2}{d\lambda^2}\mathcal{E}(\lambda\nu_1 + (1-\lambda)\nu_2) = 2\mathcal{E}(\nu_1 - \nu_2) \geq 0.$$

This shows that $\mathcal{E}$ is convex. To show strict convexity, we will show that the only compactly supported finite signed measure ν of total mass 0 on $\mathbb{C}$ with $\mathcal{E}(\nu) = 0$ is the zero measure; this in particular implies that $\mathcal{E}(\nu_1 - \nu_2) > 0$ above, since ν_1 and ν_2 are distict.

Since $\nu(\mathbb{C}) = 0$, if $0 < \epsilon < R < \infty$, then

$$\begin{aligned}-&\int_\epsilon^R \left(\iint \frac{1}{t + |z-w|} d\nu(z)d\nu(w) \right) dt \\ &= \int_\epsilon^R \left(\iint \left(\frac{1}{1+t} - \frac{1}{t+|z-w|} \right) d\nu(z)d\nu(w) \right) dt \\ &= \iint \left(\log(\epsilon + |z-w|) + \log\left(\frac{1+R}{R+|z-w|} \right) - \log(1+\epsilon) \right) d\nu(z)d\nu(w).\end{aligned}$$

We have shown above that $(z, w) \mapsto \frac{1}{t+|z-w|}$ is a positive definite kernel for each t, so that

$$\iint \frac{1}{t+|z-w|} d\nu(z) d\nu(w) \geq 0$$

for each t. The monotone convergence theorem then justifies taking the limit above as $\epsilon \to 0$ and $R \to \infty$:

$$\lim_{\epsilon \to 0} \lim_{R \to \infty} \int_{\epsilon}^{R} \left(\iint \frac{1}{t+|z-w|} d\nu(z) d\nu(w) \right) dt$$
$$= \int_{0}^{\infty} \left(\iint \frac{1}{t+|z-w|} d\nu(z) d\nu(w) \right) dt.$$

On the other hand, we have argued above that

$$\lim_{\epsilon \to 0} \iint \log(\epsilon + |z-w|) d\nu(z) d\nu(w) = \mathcal{E}(\nu) (= 0),$$

and

$$\iint \log(1+\epsilon) d\nu(z) d\nu(w) = 0$$

for all $\epsilon > 0$, since $\nu(\mathbb{C}) = 0$. Finally,

$$\lim_{R \to \infty} \iint \log\left(\frac{1+R}{R+|z-w|} \right) d\nu(z) d\nu(w) = 0$$

by the dominated convergence theorem, since ν is assumed to be compactly supported. We thus have that

$$\int_{0}^{\infty} \left(\iint \frac{1}{t+|z-w|} d\nu(z) d\nu(w) \right) dt = 0,$$

and because the inner double integral is a nonnegative function of t, this means that

$$\iint \frac{1}{t+|z-w|} d\nu(z) d\nu(w) = 0$$

for all $t > 0$.

Write

$$\frac{1}{t+|z-w|} = \frac{1}{t} \sum_{n=0}^{\infty} \left(-\frac{|z-w|}{t} \right)^n = s \sum_{n=0}^{\infty} (-s|z-w|)^n,$$

with $s = \frac{1}{t}$. We have from above that

$$\iint \sum_{n=0}^{\infty} (-s|z-w|)^n d\nu(z) d\nu(w) = 0$$

for each s. The $n = 0$ term integrates to 0 on its own, since $\nu(\mathbb{C}) = 0$. Since ν is compactly supported, given $\epsilon > 0$, we can choose $s = \frac{\epsilon}{\operatorname{diam}(\operatorname{supp}(\nu))}$ so that $\sum_{n=2}^{\infty} (s|z-w|)^n < \frac{\epsilon^2}{1-\epsilon}$. But then

$$\begin{aligned}
&\left| \iint |z-w| d\nu(z) d\nu(w) \right| \\
&\quad = \left| \frac{1}{s} \iint \sum_{n=2}^{\infty} (-s|z-w|)^n d\nu(z) d\nu(w) \right| < \frac{\epsilon \operatorname{diam}(\operatorname{supp}(\nu)) (|\nu|(\mathbb{C}))^2}{1-\epsilon};
\end{aligned}$$

i.e., $\iint |z-w| d\nu(z) d\nu(w) = 0$. Iterating this argument gives that

$$\iint |z-w|^n d\nu(z) d\nu(w) = 0$$

for all $n \in \mathbb{N}$. In particular, considering even powers gives that for all $n \geq 0$,

$$\begin{aligned}
0 &= \iint (z-w)^n (\overline{z} - \overline{w})^n d\nu(z) d\nu(w) \\
&= \sum_{j,k=0}^{n} \binom{n}{j} \binom{n}{k} \left(\int z^j \overline{z}^k d\nu(z) \right) \left(\int (-w)^{n-j} (-\overline{w})^{n-k} d\nu(w) \right)
\end{aligned}$$

Exercise 4.21 Prove by induction that this last equality implies that $\iint z^j \overline{z}^k d\nu(z) = 0$ for all j, k.

Since ν is compactly supported, the monomials $\{z^j \overline{z}^k\}$ span a dense subset of the continuous functions on the support of ν, and so the result of the exercise shows that $\nu = 0$.

All that remains is to show that $\mathcal{E}$ is lower semicontinuous. Let

$$F(\zeta, \eta) := -\log |\zeta - \eta|$$

and for $\alpha > 0$, let

$$F_\alpha(\zeta, \eta) := \min\{F(\zeta, \eta), \alpha\}.$$

Then F_α is a bounded, continuous function on $\mathbb{S}^1 \times \mathbb{S}^1$. Given a bounded continuous function g, the mapping

$$\mu \mapsto \int g d\mu$$

is continuous by definition, and the mapping $\mu \mapsto \mu \times \mu$ is also continuous, so

$$\mu \mapsto \iint_{\mathbb{S}^1 \times \mathbb{S}^1} F_\alpha(\zeta,\eta) d\mu(\zeta) d\mu(\eta)$$

is continuous on $\mathcal{P}(\mathbb{S}^1)$. By the monotone convergence theorem,

$$\mathcal{E}(\mu) = \iint F(\zeta,\eta) d\mu(\zeta) d\mu(\eta) = \sup_{\alpha>0} \iint F_\alpha(\zeta,\eta) d\mu(\zeta) d\mu(\eta);$$

that is, $\mathcal{E}$ is the supremum of continuous functions and hence lower semicontinuous. □

Empirical Spectral Measures

Let μ_n denote the empirical spectral measure of a random unitary matrix $U \in \mathbb{U}(n)$. Then μ_n is a random element of the (compact) topological space $\mathcal{P}(\mathbb{S}^1)$ (with the topology of convergence in distribution). Let $\mathbb{P}_n$ denote the distribution of μ_n in $\mathcal{P}(\mathbb{S}^1)$; it is for these $\mathbb{P}_n$ that the LDP holds, as follows.

Theorem 4.22 (Hiai–Petz) *Let $\mathbb{P}_n$ denote the distribution of the empirical spectral measure of a Haar-distributed random unitary matrix in $\mathbb{U}(n)$. The sequence $\{\mathbb{P}_n\}_{n\in\mathbb{N}}$ of measures on $\mathcal{P}(\mathbb{S}^1)$ as defined above satisfies an LDP with speed n^2 and rate function given by the logarithmic energy $\mathcal{E}$.*

Proof It is a consequence of Alaoglu's theorem that $\mathcal{P}(\mathbb{S}^1)$ is compact in the weak* topology, and so the existence of a weak LDP is the same as the full LDP. We thus proceed via Theorem 4.19: we show that

$$\mathcal{E}(\mu) \geq \sup_{A \in \mathcal{A}:\mu\in A} \left[-\liminf_{n\to\infty} \frac{1}{n^2} \log \mathbb{P}_n(A) \right] \tag{4.13}$$

and

$$\mathcal{E}(\mu) \leq \sup_{A \in \mathcal{A}:\mu\in A} \left[-\limsup_{n\to\infty} \frac{1}{n^2} \log \mathbb{P}_n(A) \right]. \tag{4.14}$$

As above, let

$$F(\zeta,\eta) := -\log|\zeta - \eta|$$

and for $\alpha > 0$, let

$$F_\alpha(\zeta,\eta) := \min\{F(\zeta,\eta), \alpha\},$$

so that

$$\mathcal{E}(\mu) = \iint F(\zeta,\eta)d\mu(\zeta)d\mu(\eta) = \sup_{\alpha>0} \iint F_\alpha(\zeta,\eta)d\mu(\zeta)d\mu(\eta).$$

Given a vector $\phi = (\phi_1, \dots, \phi_n) \in [0, 2\pi]^n$, let

$$\mu_\phi = \frac{1}{n}\sum_{j=1}^{n} \delta_{e^{i\phi_j}}.$$

Let $\mu \in \mathcal{P}(\mathbb{S}^1)$ and let A be a neighborhood of μ. Define

$$A_0 := \{\phi \in [0, 2\pi]^n : \mu_\phi \in A\}.$$

Then by the Weyl integration formula (Theorem 3.1), for any $\alpha > 0$,

$$\begin{aligned}
\mathbb{P}_n(A) &= \frac{1}{(2\pi)^n n!} \int \cdots \int_{A_0} \prod_{1\leq j<k\leq n} |e^{i\phi_j} - e^{i\phi_k}|^2 d\phi_1 \cdots d\phi_n \\
&= \frac{1}{(2\pi)^n n!} \int \cdots \int_{A_0} \exp\left\{ -\sum_{j\neq k} F(e^{i\phi_j}, e^{i\phi_k}) \right\} d\phi_1 \cdots d\phi_n \\
&\leq \frac{1}{(2\pi)^n n!} \int \cdots \int_{A_0} \exp\left\{ -\sum_{j\neq k} F_\alpha(e^{i\phi_j}, e^{i\phi_k}) \right\} d\phi_1 \cdots d\phi_n \\
&= \frac{1}{(2\pi)^n n!} \int \cdots \int_{A_0} \exp\left\{ -n^2 \iint F_\alpha(e^{ix}, e^{iy}) d\mu_\phi(x) d\mu_\phi(y) + \alpha n \right\} \\
&\qquad \times d\phi_1 \cdots d\phi_n \\
&\leq \frac{e^{\alpha n}}{n!} \exp\left\{ -n^2 \inf_{\rho\in A} \iint F_\alpha(\zeta,\eta) d\rho(\zeta) d\rho(\eta) \right\}
\end{aligned}$$

Taking logarithms, dividing by n^2, and taking the limit superior of both sides gives that

$$\limsup_{n\to\infty} \frac{1}{n^2} \log \mathbb{P}_n(A) \leq -\inf_{\rho\in A} \iint F_\alpha(\zeta,\eta) d\rho(\zeta) d\rho(\eta),$$

so that

$$\sup_{A\in\mathcal{A},\mu\in A} \left[-\limsup_{n\to\infty} \frac{1}{n^2} \log \mathbb{P}_n(A) \right] \geq \sup_{A\in\mathcal{A},\mu\in A} \left[\inf_{\rho\in A} \iint F_\alpha(\zeta,\eta) d\rho(\zeta) d\rho(\eta) \right].$$

Recall that for each $\alpha > 0$, $\rho \mapsto \iint F_\alpha(\zeta,\eta)d\rho(\zeta)d\rho(\eta)$ is continuous, so that for any $\epsilon > 0$, we can choose A_ϵ a neighborhood of μ such that

$$\inf_{\rho \in A_\epsilon} \iint F_\alpha(\zeta,\eta)d\rho(\zeta)d\rho(\eta) \geq \iint F_\alpha(\zeta,\eta)d\mu(\zeta)d\mu(\eta) - \epsilon.$$

Letting $\epsilon \to 0$ and then $\alpha \to \infty$ shows that

$$\sup_{A \in \mathcal{A}, \mu \in A} \left[-\limsup_{n\to\infty} \frac{1}{n^2} \log \mathbb{P}_n(A) \right] \geq \iint F(\zeta,\eta)d\mu(\zeta)d\mu(\eta) = \mathcal{E}(\mu),$$

and so (4.14) is proved.

To prove (4.13), we make use of a regularization of μ via the Poisson kernel. Specifically, for $0 < r < 1$, let

$$P_r(\theta) = \frac{1-r^2}{1 - 2r\cos(\theta) + r^2}.$$

If

$$f_r(e^{i\theta}) = [P_r * \mu](e^{i\theta}) = \int_{\mathbb{S}^1} P_r(\theta - \arg(\zeta))d\mu(\zeta)$$

and μ_r is the probability measure on $\mathbb{S}^1$ with density f_r, then f_r is continuous and strictly positive, $\mu_r \Rightarrow \mu$ as $r \to 1$, and $\mathcal{E}(\mu_r) \xrightarrow{r\to 1} \mathcal{E}(\mu)$ (for details, see [51]).

Let $\delta > 0$ be such that $\delta \leq f_r(z) \leq \delta^{-1}$ for $z \in \mathbb{S}^1$. Since f_r is strictly positive, the function

$$\theta \mapsto \frac{1}{2\pi} \int_0^\theta f_r(e^{it})dt$$

is invertible; let $\varphi : [0,1] \to [0,2\pi]$ denote the inverse. Then for each $n \in \mathbb{N}$ and $j \in \{1, \ldots, n\}$, define

$$0 = b_0^{(n)} < a_1^{(n)} < b_1^{(n)} < \cdots < a_n^{(n)} < b_n^{(n)} = 2\pi$$

by

$$a_j^{(n)} = \varphi\left(\frac{j - \frac{1}{2}}{n}\right) \qquad b_j^{(n)} = \varphi\left(\frac{j}{n}\right),$$

and note that this implies that for all $j = 1, \ldots, n$,

$$\frac{\pi\delta}{n} \leq \left(b_j^{(n)} - a_j^{(n)}\right) \leq \frac{\pi}{n\delta} \qquad \frac{\pi\delta}{n} \leq \left(a_j^{(n)} - b_{j-1}^{(n)}\right) \leq \frac{\pi}{n\delta}.$$

Let

$$\Delta_n := \left\{(\theta_1, \ldots, \theta_n) : a_j^{(n)} \leq \theta_j \leq b_j^{(n)}, 1 \leq j \leq n\right\}$$

and suppose that $\phi = (\phi_1, \ldots, \phi_n) \in \Delta_n$. Let $g : \mathbb{S}^1 \to \mathbb{R}$ have $\|g\|_{BL} \leq 1$ (recall that $\|g\|_{BL}$ is the maximum of the supremum norm and the Lipschitz constant of g). Then

$$
\begin{aligned}
&\left| \int_{\mathbb{S}^1} g(z) d\mu_\phi(z) - \int_{\mathbb{S}^1} g(z) d\mu_r(z) \right| \\
&= \left| \frac{1}{n} \sum_{j=1}^{n} g(e^{i\phi_j}) - \frac{1}{2\pi} \sum_{j=1}^{n} \int_{b_{j-1}^{(n)}}^{b_j^{(n)}} g(e^{i\theta}) f_r(e^{i\theta}) d\theta \right| \\
&= \left| \sum_{j=1}^{n} \frac{1}{2\pi} \int_{b_{j-1}^{(n)}}^{b_j^{(n)}} \left(g(e^{i\phi_j}) - g(e^{i\theta}) \right) f_r(e^{i\theta}) d\theta \right| \\
&\leq \max_{1 \leq j \leq n} \left| b_j^{(n)} - b_{j-1}^{(n)} \right| \\
&\leq \frac{2\pi}{n\delta}.
\end{aligned}
$$

Since the bounded-Lipschitz distance is a metric for the topology of convergence in distribution, this means that for any neighborhood A of μ_r, for n large enough,

$$
\Delta_n \subseteq A_0 = \{\phi : \mu_\phi \in A\}.
$$

Writing

$$
m_{jk}^{(n)} := \min \left\{ |e^{is} - e^{it}| : a_j^{(n)} \leq s \leq b_j^{(n)}, a_k^{(n)} \leq t \leq b_k^{(n)} \right\},
$$

it now follows from the Weyl integration formula that

$$
\begin{aligned}
\mathbb{P}_n(A) &= \frac{1}{(2\pi)^n n!} \int \cdots \int_{A_0} \prod_{1 \leq j < k \leq n} |e^{i\phi_j} - e^{i\phi_k}|^2 d\phi_1 \cdots d\phi_n \\
&\geq \frac{1}{(2\pi)^n n!} \int \cdots \int_{\Delta_n} \prod_{1 \leq j < k \leq n} |e^{i\phi_j} - e^{i\phi_k}|^2 d\phi_1 \cdots d\phi_n \\
&\geq \frac{1}{(2\pi)^n n!} \prod_{1 \leq j < k \leq n} \left[m_{jk}^{(n)} \right]^2 \int_{a_1^{(n)}}^{b_1^{(n)}} \cdots \int_{a_n^{(n)}}^{b_n^{(n)}} d\phi_1 \cdots d\phi_n \\
&\geq \frac{1}{n!} \left(\frac{\delta}{2n} \right)^n \prod_{1 \leq j < k \leq n} \left[m_{jk}^{(n)} \right]^2,
\end{aligned}
$$

since $b_n^{(n)} - a_j^{(n)} \geq \frac{\pi\delta}{n}$. Taking logarithms, diving by n^2, and taking limits inferior thus gives

$$\liminf_{n\to\infty}\frac{1}{n^2}\log\mathbb{P}_n(A)\geq\liminf_{n\to\infty}\frac{2}{n^2}\sum_{1\leq j<j\leq n}\log(m_{jk}^{(n)}).$$

Since φ is increasing with $a_j^{(n)}=\frac{j-\frac{1}{2}}{n}$ and $b_j^{(n)}=\frac{j}{n}$,

$$\begin{aligned}\lim_{n\to\infty}\frac{2}{n^2}\sum_{1\leq j<k\leq n}\log(m_{jk}^{(n)})&=\lim_{n\to\infty}\left[\frac{2}{n^2}\sum_{1\leq j<k\leq n}\log\left(\min_{\substack{\frac{j-\frac{1}{2}}{n}\leq u\leq\frac{j}{n}\\ \frac{k-\frac{1}{2}}{n}\leq v\leq\frac{k}{n}}}\left|e^{i\varphi(u)}-e^{i\varphi(v)}\right|\right)\right]\\&=2\iint_{0\leq u<v\leq 1}\log\left|e^{i\varphi(u)}-e^{i\varphi(v)}\right|dudv\\&=\frac{1}{(2\pi)^2}\int_0^{2\pi}\int_0^{2\pi}f_r(e^{is})f_r(e^{it})\log|e^{is}-e^{it}|dsdt\\&=-\mathcal{E}(\mu_r),\end{aligned}$$

where the convergence of the Riemann sums to the integral is valid because the integrand is bounded below by $\log\left(\frac{\delta}{n}\right)$.

We thus have that for any neighborhood A of μ_r,

$$-\liminf_{n\to\infty}\frac{1}{n^2}\log\mathbb{P}_n(A)\leq\mathcal{E}(\mu_r).$$

Since $\mu_r\Rightarrow\mu$, this also holds for any neighborhood A of μ, for r close enough to 1, so that

$$\sup_{A\in\mathcal{A},\mu\in A}\left[-\liminf_{n\to\infty}\frac{1}{n^2}\log\mathbb{P}_n(A)\right]\leq\limsup_{r\to1}\mathcal{E}(\mu_r)=\mathcal{E}(\mu).$$

This completes the proof of (4.13). □

Notes and References

The paper [55] of Hough–Krishnapur–Peres–Virág gives a beautiful survey on determinantal point processes with many applications; this paper is in particular the source of Theorem 4.1 and its proof.

Theorem 4.9 and a multivariate generalization were proved by Wieand [106], by exploiting the connection with Toeplitz matrices. She showed the following.

Theorem 4.23 (Wieand) *Let U be a Haar-distributed random matrix in $\mathbb{U}(n)$. For $0\leq\alpha<\beta<2\pi$, let $\mathcal{N}_{\alpha,\beta}^{(n)}$ denote the number of eigenangles of U lying in $[\alpha,\beta]$. The finite-dimensional distributions of the process*

$$\frac{\pi}{\sqrt{\log(n)}}\left(\mathcal{N}_{\alpha,\beta}^{(n)} - \mathbb{E}\mathcal{N}_{\alpha,\beta}^{(n)}\right)_{0\le\alpha<\beta<2\pi}$$

converge to those of a centered Gaussian process $\{Z_{\alpha,\beta}\}_{0\le\alpha<\beta<2\pi}$ *with covariance*

$$\mathbb{E}\left[Z_{\alpha,\beta}Z_{\alpha',\beta'}\right] = \begin{cases} 1, & \alpha=\alpha',\beta=\beta'; \\ \frac{1}{2}, & \alpha=\alpha',\beta\neq\beta'; \\ \frac{1}{2}, & \alpha\neq\alpha',\beta=\beta'; \\ -\frac{1}{2}, & \beta=\alpha'; \\ 0, & \textit{otherwise}. \end{cases}$$

These surprising correlations have been the result of further study, in particular by Diaconis and Evans [32] and Hughes, Keating, and O'Connell [56]. In [95], Soshnikov proved univariate central limit theorems for local and global statistics of eigenvalues of random matrices from the classical compact groups.

Fast rates of convergence in the univariate central limit theorem for $\mathrm{Tr}(U^k)$ for k fixed and U distributed according to Haar measure were found for $U \in \mathbb{O}(n)$ by Stein [96] and Johansson [60]. Johansson also found rates of convergence in the unitary and symplectic cases; his results are as follows.

Theorem 4.24 (Johansson [60]) *Let* $U \in G(n)$ *where* $G(n)$ *is one of* $\mathbb{O}(n)$, $\mathbb{U}(n)$, *and* $\mathbb{S}\mathrm{p}(2n)$. *Let* $k \in \mathbb{N}$ *and let*

$$X_k = \frac{1}{\sqrt{k}}\left(\mathrm{Tr}(U^k) - \mathbb{E}\,\mathrm{Tr}(U^k)\right).$$

There are positive constants C_i *and* δ_i $(1 \le i \le 3)$, *independent of* n, *such that the following hold.*

1. *For* U *distributed according to Haar measure on* $\mathbb{U}(n)$, *and* Z *a standard complex Gaussian random variable,*

$$d_{TV}(X_k, Z) \le C_1 n^{-\delta_1 n}.$$

2. *For* U *distributed according to Haar measure on* $\mathbb{O}(n)$, *and* Z *a standard (real) Gaussian random variable,*

$$d_{TV}(X, Z) \le C_2 e^{-\delta_2 n}.$$

3. *For* U *distributed according to Haar measure on* $\mathbb{S}\mathrm{p}(2n)$, *and* Z *a standard complex Gaussian random variable,*

$$d_{TV}(X, Z) \le C_3 e^{-\delta_3 n}.$$

Rates of convergence in the multivariate case were obtained by Döbler and Stolz [38] via Stein's method, following work of Fulman [46].

In the paper [32] of Diaconis and Evans, they did further work to expand the class of test functions from $H_2^{\frac{1}{2}}$ so as to prove multivariate limit theorems for the number of eigenvalues in an arc, which in particular recovered Wieand's result above.

For a general introduction to the theory of large deviations, see the book [30] of Dembo and Zeitouni. The large deviations principle for the empirical spectral measure (Theorem 4.22) is due to Hiai and Petz [51] (see also their book [52]), and we have followed their proof fairly closely, with background on the logarithmic energy taken from [10] and [69].

5

Concentration of Measure

5.1 The Concentration of Measure Phenomenon

The phenomenon of concentration of measure arises frequently as a tool in probability and related areas; following its use by Vitali Milman in his probabilistic proof of Dvoretzky's theorem, the explicit study and more systematic exploitation of this phenomenon has become a large and influential area. The basic idea is that in various settings, functions with small local fluctuations are often essentially constant, where "essentially" should be interpreted in the probabilistic sense that with high probability, such functions are close to their means.

The following result in classical probability gives a first example of a concentration phenomenon.

Theorem 5.1 (Bernstein's inequality) *Let $\{X_j\}_{j=1}^n$ be independent random variables such that, for each j, $|X_j| \leq 1$ almost surely. Let $S_n := \sum_{j=1}^n X_j$ and let $\sigma^2 = \mathrm{Var}\left(\sum_{j=1}^n X_j\right)$. Then for all $t > 0$,*

$$\mathbb{P}\left[\left|\frac{S_n}{n} - \mathbb{E}\left[\frac{S_n}{n}\right]\right| > t\right] \leq C\exp\left(-\min\left\{\frac{n^2t^2}{2\sigma^2}, \frac{nt}{2}\right\}\right).$$

Letting $t = \frac{A\sigma^2}{n}$ for a large constant A gives that

$$\mathbb{P}\left[\left|\frac{S_n}{n} - \mathbb{E}\left[\frac{S_n}{n}\right]\right| > \frac{A\sigma^2}{n}\right] \leq Ce^{-\frac{A\sigma^2}{2}}.$$

That is, if n is large, it is likely that the average of n bounded independent random variables is within a window of size $O\left(\frac{\sigma^2}{n}\right)$ about its mean. It is reasonable to think of the average of n random variables as a statistic with small local fluctuations, since if just the value of one (or a few) of the random variables is changed, the average can only change on the scale of $\frac{1}{n}$.

Another classical appearance of concentration of measure is that of Gaussian measure concentration:

Theorem 5.2 *Let $f : \mathbb{R}^n \to \mathbb{R}$ be Lipschitz with Lipschitz constant L, and let $Z = (Z_1, \ldots, Z_n)$ be a standard Gaussian random vector in $\mathbb{R}^n$. Let M be a median of f; i.e., $\mathbb{P}[f(Z) \geq M] \geq \frac{1}{2}$ and $\mathbb{P}[f(Z) \leq M] \geq \frac{1}{2}$. Then*

$$P\big[|f(Z) - M| \geq Lt\big] \leq 2e^{-\frac{t^2}{2}}.$$

The following statement about uniform measure on the sphere is analogous to the previous result; the difference in appearance of the exponent is only due to the choice of normalization of the random vector.

Theorem 5.3 (Lévy's lemma) *Let $f : \mathbb{S}^{n-1} \to \mathbb{R}$ be Lipschitz with Lipschitz constant L, and let X be a uniform random vector in $\mathbb{S}^{n-1}$. Let M be a median of f; that is, $\mathbb{P}[f(X) \geq M] \geq \frac{1}{2}$ and $\mathbb{P}[f(X) \leq M] \geq \frac{1}{2}$. Then*

$$\mathbb{P}\left[\big|f(X) - M\big| \geq Lt\right] \leq 2e^{-(n-2)t^2}.$$

Both results can be loosely interpreted as saying that if the local fluctuations of a function are controlled (the function is Lipschitz), then the function is essentially constant.

Concentration results are often formulated in terms of the mean rather than a median, as follows.

Corollary 5.4 *Let $f : \mathbb{S}^{n-1} \to \mathbb{R}$ be Lipschitz with Lipschitz constant L, and let X be a uniform random vector in $\mathbb{S}^{n-1}$. Then if M_f denotes a median of f with respect to uniform measure on $\mathbb{S}^{n-1}$,*

$$|\mathbb{E}f(X) - M_f| \leq L\sqrt{\frac{\pi}{n-2}}$$

and

$$\mathbb{P}[|f(X) - \mathbb{E}f(X)| \geq Lt] \leq e^{\pi - \frac{nt^2}{4}}.$$

Proof First note that Lévy's lemma and Fubini's theorem imply that

$$\begin{aligned}\big|\mathbb{E}f(X) - M_f\big| &\leq \mathbb{E}\big|f(X) - M_f\big| \\ &= \int_0^\infty \mathbb{P}\Big[\big|f(X) - M_f\big| > t\Big]dt \leq \int_0^\infty 2e^{-\frac{(n-2)t^2}{L^2}}dt = L\sqrt{\frac{\pi}{n-2}}.\end{aligned}$$

If $t > 2\sqrt{\frac{\pi}{n-2}}$, then

$$\begin{aligned}\mathbb{P}\left[|f(X) - \mathbb{E}f(X)| > Lt\right] &\leq \mathbb{P}\left[\left|f(X) - M_f\right| > Lt - \left|M_f - \mathbb{E}f(X)\right|\right] \\ &\leq \mathbb{P}\left[\left|f(X) - M_f\right| > L\left(t - \sqrt{\frac{\pi}{n-2}}\right)\right] \\ &\leq 2e^{-\frac{(n-2)t^2}{4}}.\end{aligned}$$

On the other hand, if $t \leq 2\sqrt{\frac{\pi}{n-2}}$, then

$$e^{\pi - \frac{(n-2)t^2}{4}} \geq 1,$$

so the statement holds trivially. □

5.2 Logarithmic Sobolev Inequalities and Concentration

Knowing that a metric probability space possesses a concentration of measure property along the lines of Lévy's lemma opens many doors; however, it is not a priori clear how to show that such a property holds or to determine what the optimal (or even good) constants are. In this section we discuss obtaining measure concentration via logarithmic Sobolev inequalities.

In what follows let (X, d) be a metric space equipped with a Borel probability measure $\mathbb{P}$, with $\mathbb{E}$ denoting expectation with respect to $\mathbb{P}$. The **entropy** of a measurable function $f : X \to [0, \infty)$ with respect to $\mathbb{P}$ is

$$\mathrm{Ent}(f) := \mathbb{E}\big[f \log(f)\big] - (\mathbb{E}f)\log(\mathbb{E}f). \tag{5.1}$$

For $c > 0$, $\mathrm{Ent}(cf) = c\,\mathrm{Ent}(f)$, and it follows from Jensen's inequality that $\mathrm{Ent}(f) \geq 0$.

Since X is an arbitrary metric measure space, it may not have a smooth structure. Nevertheless, the concept of the *length* of the gradient extends, as follows. A function $g : X \to \mathbb{R}$ is locally Lipschitz if for all $x \in X$, there is a neighborhood $U \subseteq X$ of x on which g is Lipschitz; for a locally Lipschitz function $g : X \to \mathbb{R}$, the length of the gradient of g at x is defined by

$$|\nabla g|(x) := \limsup_{y \to x} \frac{|g(y) - g(x)|}{d(y, x)}.$$

For smooth functions $\phi : \mathbb{R} \to \mathbb{R}$, this length of gradient satisfies the chain rule:

$$|\nabla \phi(f)| \leq |\phi'(f)||\nabla f|.$$

Definition The space $(X, d, \mathbb{P})$ satisfies a **logarithmic Sobolev inequality** (or **log-Sobolev inequality** or **LSI**) with constant $C > 0$ if, for every locally Lipschitz $f : X \to \mathbb{R}$,

$$\mathrm{Ent}(f^2) \leq 2C\mathbb{E}\left(|\nabla f|^2\right). \tag{5.2}$$

The reason for our interest in log-Sobolev inequalities is that they imply measure concentration for Lipschitz functions, via what is known as the Herbst argument.

Theorem 5.5 *Suppose that $(X, d, \mathbb{P})$ satisfies a log-Sobolev inequality with constant $C > 0$. Then if $F : X \to \mathbb{R}$ is 1-Lipschitz, $\mathbb{E}|F| < \infty$, and for every $r \geq 0$,*

$$\mathbb{P}\Big[\big|F - \mathbb{E}F\big| \geq r\Big] \leq 2e^{-r^2/2C}.$$

Proof (the Herbst argument) First consider the case that F is bounded as well as Lipschitz, and note also that by replacing F with $F - \mathbb{E}F$, it suffices to treat the case $\mathbb{E}F = 0$. For $\lambda > 0$, it follows by Markov's inequality that

$$\mathbb{P}\left[F \geq r\right] = \mathbb{P}\left[e^{\lambda F} \geq e^{\lambda r}\right] \leq e^{-\lambda r}\mathbb{E}e^{\lambda F}. \tag{5.3}$$

For notational convenience, let $H(\lambda) := \mathbb{E}e^{\lambda F}$, and consider the function f with

$$f^2 := e^{\lambda F}.$$

Then

$$\mathrm{Ent}(f^2) = \mathbb{E}\left[\lambda F e^{\lambda F}\right] - H(\lambda)\log H(\lambda),$$

and

$$|\nabla f(x)| \leq e^{\frac{\lambda F(x)}{2}}\left(\frac{\lambda}{2}\right)|\nabla F(x)|.$$

Taking expectation and using the fact that F is 1-Lipschitz (so that $|\nabla F| \leq 1$) gives

$$\mathbb{E}|\nabla f|^2 \leq \frac{\lambda^2}{4}\mathbb{E}\left[|\nabla F|^2 e^{\lambda F}\right] \leq \frac{\lambda^2}{4}\mathbb{E}\left[e^{\lambda F}\right] = \frac{\lambda^2}{4}H(\lambda);$$

applying the LSI with constant C to this f thus yields

$$\mathbb{E}\left[\lambda F e^{\lambda F}\right] - H(\lambda)\log H(\lambda) = \lambda H'(\lambda) - H(\lambda)\log H(\lambda) \leq \frac{C\lambda^2}{2}H(\lambda),$$

or rearranging,

$$\frac{H'(\lambda)}{\lambda H(\lambda)} - \frac{\log H(\lambda)}{\lambda^2} \le \frac{C}{2}.$$

Now, if $K(\lambda) := \frac{\log H(\lambda)}{\lambda}$, then the right-hand side is just $K'(\lambda)$, and so we have the simple differential inequality

$$K'(\lambda) \le \frac{C}{2}.$$

Since $H(0) = 1$,

$$\lim_{\lambda \to 0} K(\lambda) = \lim_{\lambda \to 0} \frac{H'(\lambda)}{H(\lambda)} = \lim_{\lambda \to 0} \frac{\mathbb{E}\left[Fe^{\lambda F}\right]}{\mathbb{E}\left[e^{\lambda F}\right]} = \mathbb{E}F = 0,$$

and thus

$$K(\lambda) = \int_0^\lambda K'(s)ds \le \int_0^\lambda \frac{C}{2} ds = \frac{C\lambda}{2}.$$

In other words,

$$\mathbb{E}\left[e^{\lambda F}\right] = H(\lambda) = e^{\lambda K(\lambda)} \le e^{\frac{C\lambda^2}{2}}.$$

It follows from this last estimate together with (5.3) that for $F : X \to \mathbb{R}$ that is 1-Lipschitz and bounded,

$$\mathbb{P}\left[F \ge \mathbb{E}F + r\right] \le e^{-\lambda r + \frac{C\lambda^2}{2}}.$$

Choosing $\lambda = \frac{r}{C}$ completes the proof under the assumption that F is bounded.

In the general case, let $\epsilon > 0$ and define the truncation F_ϵ by

$$F_\epsilon(x) := \begin{cases} -\frac{1}{\epsilon}, & F(x) \le -\frac{1}{\epsilon}; \\ F(x), & -\frac{1}{\epsilon} \le F(x) \le \frac{1}{\epsilon}; \\ \frac{1}{\epsilon}, & F(x) \ge \frac{1}{\epsilon}. \end{cases}$$

Then F_ϵ is 1-Lipschitz and bounded so that by the argument above,

$$\mathbb{E}\left[e^{\lambda F_\epsilon}\right] \le e^{\lambda \mathbb{E}F_\epsilon + \frac{C\lambda^2}{2}}.$$

The truncation F_ϵ approaches F pointwise as $\epsilon \to 0$, so by Fatou's lemma,

$$\mathbb{E}\left[e^{\lambda F}\right] \le e^{\liminf_{\epsilon \to 0} \lambda \mathbb{E}F_\epsilon} e^{\frac{C\lambda^2}{2}}.$$

It remains to show that $\mathbb{E}F_\epsilon \xrightarrow{\epsilon \to 0} \mathbb{E}F$ (in particular that $\mathbb{E}F$ is defined); the proof is then completed exactly as in the bounded case.

Now, F takes on some finite value at each point in X, so there is a constant K such that

$$\mathbb{P}\left[|F| \leq K\right] \geq \frac{7}{8};$$

moreover, F_ϵ converges pointwise, hence also in probability to F, so there is some ϵ_o such that for $\epsilon < \epsilon_o$,

$$\mathbb{P}\left[|F_\epsilon - F| > K\right] < \frac{1}{8},$$

and so $\mathbb{P}\left[|F_\epsilon| \leq 2K\right] \geq \frac{3}{4}$. On the other hand, since $|F_\epsilon|$ is bounded and 1-Lipschitz, it has already been shown that

$$\mathbb{P}\left[\Big||F_\epsilon| - \mathbb{E}|F_\epsilon|\Big| > t\right] \leq 2e^{-\frac{t^2}{2C}}, \tag{5.4}$$

so that there is some r (depending only on C) such that

$$\mathbb{P}\left[\Big||F_\epsilon| - \mathbb{E}|F_\epsilon|\Big| > r\right] \leq \frac{1}{4}.$$

It follows that for $\epsilon < \epsilon_o$, the set

$$\{|F_\epsilon| < 2K\} \cap \left\{\Big||F_\epsilon| - \mathbb{E}|F_\epsilon|\Big| \leq r\right\}$$

has probability at least $\frac{1}{2}$, and is in particular non-empty. But on this set, $\mathbb{E}|F_\epsilon| \leq 2K + r$, and so $\mathbb{E}|F_\epsilon|$ is uniformly bounded, independent of ϵ.

It follows from the version of (5.4) for F_ϵ itself and Fubini's theorem that

$$\mathbb{E}\left|F_\epsilon - \mathbb{E}F_\epsilon\right|^2 = \int_0^\infty t\mathbb{P}\left[|F_\epsilon - \mathbb{E}F_\epsilon| > t\right] dt \leq \int_0^\infty 2te^{-\frac{t^2}{2C}} dt = 2C,$$

and using Fatou's lemma again gives that

$$\mathbb{E}|F - \mathbb{E}F_\epsilon|^2 \leq \liminf_{\epsilon \to 0} \mathbb{E}|F_\epsilon - \mathbb{E}F_\epsilon|^2 \leq 2C.$$

In particular

$$\mathbb{E}|F| \leq \mathbb{E}|F - \mathbb{E}F_\epsilon| + \mathbb{E}|F_\epsilon| \leq \sqrt{2C} + 2K + r,$$

and so F is integrable. One final application of the convergence of F_ϵ to F in probability gives

$$\begin{aligned}|\mathbb{E}F - \mathbb{E}F_\epsilon| &\leq \delta + \mathbb{E}|F_\epsilon - F|\mathbb{1}_{|F_\epsilon - F| > \delta} \\ &\leq \delta + \sqrt{\mathbb{E}|F_\epsilon - F|^2\mathbb{P}\left[|F_\epsilon - F| > \delta\right]} \xrightarrow{\epsilon \to 0} 0.\end{aligned}$$

□

Log-Sobolev inequalities can be transfered between spaces via Lipschitz maps, as follows.

Lemma 5.6 *Let $(X, d_X, \mathbb{P})$ be a metric space equipped with a Borel probability measure $\mathbb{P}$ and let (Y, d_Y) be a metric space. Suppose that $(X, d_X, \mathbb{P})$ satisfies a log-Sobolev inequality with constant C. Let $F : X \to Y$ be a Lipschitz map with Lipschitz constant L, and let $\mathbb{P}_F$ be the push-forward of $\mathbb{P}$ to Y via F; i.e, if $A \subseteq Y$ is a Borel set, then*

$$\mathbb{P}_F(A) = \mathbb{P}(F^{-1}(A)).$$

Then $(Y, d_Y, \mathbb{P}_F)$ satisfies a log-Sobolev inequality with constant CL^2.

Proof Let $g : Y \to \mathbb{R}$ be locally Lipschitz. Then $g \circ F : X \to \mathbb{R}$ is locally Lipschitz, and at each $x \in X$,

$$|\nabla(g \circ F)|(x) \leq L|\nabla g|(F(x)).$$

Applying the LSI on X to $g \circ F$ thus yields

$$\begin{aligned}
\mathrm{Ent}_{\mathbb{P}_F}(g) &= \mathrm{Ent}_{\mathbb{P}}(g \circ F) \\
&\leq 2C \int \left|\nabla g \circ F\right|^2(x) d\mathbb{P}(x) \\
&\leq 2CL^2 \int \left|\nabla F\right|^2(F(x)) d\mathbb{P}(x) = 2CL^2 \int \left|\nabla F\right|^2(y) d\mathbb{P}_F(y).
\end{aligned}$$

□

A key advantage of the approach to concentration via log-Sobolev inequalities is that log-Sobolev inequalities *tensorize*; that is, if one has the same LSI on each of some finite collection of spaces, *the same* LSI holds on the product space, independent of the number of factors, as follows.

Theorem 5.7 *Suppose that each of the metric probability spaces (X_i, d_i, μ_i) $(1 \leq i \leq n)$ satisifes a log-Sobolev inequality: for each i there is a $C_i > 0$ such that for every locally Lipschitz function $f : X_i \to \mathbb{R}$,*

$$\mathrm{Ent}_{\mu_i}(f^2) \leq 2C_i \int |\nabla_{X_i} f|^2 d\mu_i.$$

Let $X = X_1 \times \cdots \times X_n$ equipped with the product probability measure $\mu := \mu_1 \otimes \cdots \otimes \mu_n$. Then for every locally Lipschitz function $f : X \to \mathbb{R}$,

$$\mathrm{Ent}_\mu(f^2) \leq 2C \int \sum_{i=1}^{n} |\nabla_{X_i} f|^2 d\mu,$$

where

$$|\nabla_{X_i} f(x_1,\ldots,x_n)| = \limsup_{y_i \to x_i} \frac{|f(x_1,\ldots,x_{i-1},y_i,x_{i+1},\ldots,x_n) - f(x_1,\ldots,x_n)|}{d_i(y_i,x_i)}$$

and $C = \max_{1 \le i \le n} C_i$.

The crucial point here is that the constant C does not get worse with the number of factors; that is, the lemma gives *dimension-free* tensorization of log-Sobolev inequalities.

The theorem follows immediately from the following property of entropy.

Proposition 5.8 *Let* $X = X_1 \times \cdots \times X_n$ *and* $\mu = \mu_1 \otimes \cdots \otimes \mu_n$ *as above, and suppose that* $f : X \to [0,\infty)$. *For* $\{x_j\}_{1 \le j \le n, j \ne i}$ *fixed, write*

$$f_i(x_i) = f(x_1,\ldots,x_n),$$

thought of as a function of x_i. *Then*

$$\mathrm{Ent}_\mu(f) \le \sum_{i=1}^{n} \int \mathrm{Ent}_{\mu_i}(f_i) d\mu.$$

Proof The proof makes use of the following dual formulation of the definition of entropy: given a probability space $(\Omega, \mathcal{F}, \mathbb{P})$, the definition of $\mathrm{Ent}(f) = \mathrm{Ent}_{\mathbb{P}}(f)$ given in Equation (5.1) is equivalent to

$$\mathrm{Ent}_{\mathbb{P}}(f) := \sup\left\{ \int fg d\mathbb{P} \middle| \int e^g d\mathbb{P} \le 1 \right\},$$

as follows.

First, for simplicity we may assume that $\int f d\mathbb{P} = 1$, since both expressions for the entropy are homogeneous of degree 1. Then the expression in (5.1) is

$$\mathrm{Ent}_{\mathbb{P}}(f) = \int f \log(f) d\mathbb{P}.$$

Now, if $g := \log(f)$, then $\int e^g d\mathbb{P} = \int f d\mathbb{P} = 1$, and so

$$\int f \log(f) d\mathbb{P} = \int fg d\mathbb{P} \le \sup\left\{ \int fg d\mathbb{P} \middle| \int e^g d\mathbb{P} \le 1 \right\}.$$

On the other hand, Young's inequality says that for $u \ge 0$ and $v \in \mathbb{R}$,

$$uv \le u \log(u) - u + e^v;$$

applying this to $u = f$ and $v = g$ and integrating shows that

$$\sup\left\{ \int fg d\mathbb{P} \middle| \int e^g d\mathbb{P} \le 1 \right\} \le \int f \log(f) d\mathbb{P}.$$

Working now with the alternative definition of entropy, given g such that $\int e^g d\mu \leq 1$, for each i define

$$g^i(x_1, \ldots, x_n) := \log\left(\frac{\int e^{g(y_1,\ldots,y_{i-1},x_i,\ldots,x_n)} d\mu_1(y_1)\cdots d\mu_{i-1}(y_{i-1})}{\int e^{g(y_1,\ldots,y_i,x_{i+1},\ldots,x_n)} d\mu_1(y_1)\cdots d\mu_i(y_i)}\right),$$

(note that g^i only actually depends on $x_i, \ldots, x_n$). Then

$$\sum_{i=1}^n g^i(x_1, \ldots, x_n) = \log\left(\frac{e^{g(x_1,\ldots,x_n)}}{\int e^{g(y_1,\ldots,y_n)} d\mu_1(y_1)\cdots d\mu_n(y_n)}\right) \geq g(x_1, \ldots, x_n),$$

and by construction,

$$\int e^{(g^i)_i} d\mu_i = \int\left(\frac{\int e^{g(y_1,\ldots,y_{i-1},x_i,\ldots,x_n)} d\mu_1(y_1)\cdots d\mu_{i-1}(y_{i-1})}{\int e^{g(y_1,\ldots,y_i,x_{i+1},\ldots,x_n)} d\mu_1(y_1)\cdots d\mu_i(y_i)}\right) d\mu_i(x_i) = 1.$$

Applying these two estimates together with Fubini's theorem yields

$$\int fg d\mu \leq \sum_{i=1}^n \int fg^i d\mu = \sum_{i=1}^n \int \left(\int f_i (g^i)_i d\mu_i\right) d\mu \leq \sum_{i=1}^n \int \operatorname{Ent}_{\mu_i}(f_i) d\mu.$$

□

In general, tensorizing concentration inequalities results in a loss in the constant which gets worse with the number of factors. This is why concentration as a consequence of a log-Sobolev inequality is so valuable: product spaces have the same type of concentration phenomena as their factors, as follows.

Theorem 5.9 *Let $(X_1, d_1, \mu_1), \ldots, (X_n, d_n, \mu_n)$ be compact metric probability spaces. Suppose that for each i, (X_i, d_i, μ_i) satisfies a log-Sobolev inequality with constant C_i. Let $X = X_1 \times \cdots \times X_n$ be equipped with the product measure $\mu = \mu_1 \otimes \cdots \otimes \mu_n$ and the ℓ_2-sum metric*

$$d^2((x_1, \ldots, x_n), (y_1, \ldots, y_n)) = \sum_{i=1}^n d_i^2(x_i, y_i).$$

If $F : X \to \mathbb{R}$ is 1-Lipschitz, then for every $r \geq 0$,

$$\mathbb{P}\Big[\big|F - \mathbb{E}F\big| \geq r\Big] \leq 2e^{-r^2/4C},$$

where $C = \max_{1 \leq i \leq n} C_i$.

Proof The main point is to connect the Lipschitz condition on F to the quantity

$$\sum_{i=1}^n \big|\nabla_{X_i} F\big|^2$$

appearing in Theorem 5.7; the rest of the proof is a repeat of the Herbst argument.

Given $F : X \to \mathbb{R}$ 1-Lipschitz, for each $\epsilon > 0$ define the function

$$F_\epsilon(x) = \inf_{z\in X}\left[F(z) + \sqrt{\epsilon^2 + d^2(x,z)}\right].$$

Then for all $x \in X$,

$$F(x) \le F_\epsilon(x) \le F(x) + \epsilon$$

(the first inequality is because F is 1-Lipschitz and the second is by choosing $z = x$ in the infimum).

Fix $x = (x_1, \ldots, x_n) \in X$. Since X is compact, there is an $a \in X$ such that $F_\epsilon(x) = F(a) + \sqrt{\epsilon^2 + d^2(x,a)}$. Now let $y_i \in X_i$, and let

$$x^{(i,y_i)} = (x_1, \ldots, x_{i-1}, y_i, x_{i+1}, \ldots, x_n).$$

Then

$$\begin{aligned} F_\epsilon(x^{(i,y_i)}) - F_\epsilon(x) &\le \sqrt{\epsilon^2 + d^2(x^{(i,y_i)},a)} - \sqrt{\epsilon^2 + d^2(x,a)} \\ &= \frac{d_i^2(y_i,a_i) - d_i^2(x_i,a_i)}{\sqrt{\epsilon^2 + d^2(x^{(i,y_i)},a)} + \sqrt{\epsilon^2 + d^2(x,a)}} \\ &\le \frac{d_i(x_i,y_i)\,[d_i(a_i,y_i) + d_i(x_i,a_i)]}{\sqrt{\epsilon^2 + d^2(x^{(i,y_i)},a)} + \sqrt{\epsilon^2 + d^2(x,a)}}, \end{aligned}$$

by repeated applications of the triangle inequality. It follows that

$$\limsup_{y_i \to x_i} \frac{F_\epsilon(x^{(i,y_i)}) - F_\epsilon(x)}{d_i(x_i,y_i)} \le \frac{d_i(x_i,a_i)}{\sqrt{\epsilon^2 + d^2(x,a)}},$$

and so if $|\nabla^+_{X_i} F_\epsilon|(x) = \limsup_{y_i \to x_i} \frac{[F_\epsilon(x^{(i,y_i)}) - F_\epsilon(x)]_+}{d_i(x_i,y_i)}$, then $\sum_{i=1}^n |\nabla^+_{X_i} F_\epsilon|^2(x) \le 1$. Applying the same argument to $-F_\epsilon$ then gives that $\sum_{i=1}^n \left|\nabla_{X_i} F_\epsilon\right|^2(x) \le 2$.

At this point, one can apply the Herbst argument to F_ϵ, using the result of Theorem 5.7 and everywhere replacing $|\nabla F_\epsilon|^2(x)$ with $\sum_{i=1}^n \left|\nabla_{X_i} F_\epsilon\right|^2(x)$. From this it follows that

$$\mathbb{P}\Big[\left|F_\epsilon - \mathbb{E}F_\epsilon\right| \ge r\Big] \le 2e^{-r^2/4C},$$

with $C = \max_{1 \le i \le n} C_i$. The result now follows from the monotone convergence theorem, letting $\epsilon \to 0$. □

5.3 The Bakry–Émery Criterion and Concentration for the Classical Compact Groups

All of the classical compact matrix groups (except $\mathbb{O}(n)$, which is disconnected), satisfy a concentration of measure property similar to the one on the sphere given in Lévy's lemma. In almost every case, the optimal (i.e., with smallest constants) log-Sobolev inequalities follow from the Bakry–Émery curvature criterion. We begin with some background in Riemannian geometry.

Riemannian Geometry and Lie Groups

Let M be a smooth manifold embedded in Euclidean space. A tangent vector at a point $p \in M$ can be realized as the tangent vector $\gamma'(0)$ to some curve $\gamma : (-\epsilon, \epsilon) \to M$ with $\gamma(0) = p$:

$$\gamma'(0) = \lim_{s \to 0} \frac{\gamma(s) - p}{s},$$

where the operations are taking place in the ambient Euclidean space. On an abstract manifold, tangent vectors to a point are defined similarly as equivalence classes of curves through that point, but we will not need this in what follows. The set of tangent vectors to M at a point p is denoted T_pM and the set of all tangent vectors to M is denoted TM.

The manifolds we are working with all have the additional structure of a Riemannian metric. A **Riemannian manifold** (M, g) is a smooth manifold together with a **Riemannian metric** g; i.e., a family of inner products: at each point $p \in M$, $g_p : T_pM \times T_pM \to \mathbb{R}$ defines an inner product on the tangent space T_pM to M at p. A manifold embedded in Euclidean space inherits a metric just by restricting the ambient Euclidean metric. Different embeddings can give rise to different metrics, but for the classical groups we will stick with the canonical embedding and resulting metric.

Given a smooth function $f : M \to N$ between manifolds, the **differential** or **push-forward** of f at $p \in M$ is the map $(f_*)_p : T_pM \to T_{f(p)}N$ that is defined as follows. Given a curve $\gamma : (-\epsilon, \epsilon) \to M$ with $\gamma(0) = p$ and $\gamma'(0) = X$,

$$(f_*)_p(X) = \left. \frac{d}{dt} f(\gamma(t)) \right|_{t=0}.$$

The definition of f_* is independent of γ.

A **vector field** X on M is a smooth (infinitely differentiable) map $X : M \to TM$ such that for each $p \in M$, $X(p) \in T_pM$. Note that the push-forward f_* can then be used to define a vector field f_*X on N. From a different perspective, the definition of f_* can be turned around to give a way that a smooth vector field X on M acts as a differential operator: given a vector field X, for any smooth function f on M, the function $X(f)$ is defined by the requirement that for any curve $\gamma : (-\epsilon, \epsilon) \to M$ with $\gamma(0) = p$ and $\gamma'(0) = X(p)$,

$$X(f)(p) = \frac{d}{dt} f(\gamma(t))\bigg|_{t=0};$$

that is, $X(f)(p) = (f_*)_p(X)$.

It is sometimes convenient to work in coordinates. A **local frame** $\{L_i\}$ is a collection of vector fields defined on an open set $U \subseteq M$ such that at each point $p \in U$, the vectors $\{L_i(p)\} \subseteq T_pM$ form a basis of T_pM. The vector fields $\{L_i\}$ are called a **local orthonormal frame** if at each point in U, the $\{L_i\}$ are orthonormal with repect to g. Some manifolds only have local frames, not global ones; that is, you cannot define a smooth family of vector fields over the whole manifold that forms a basis of the tangent space at each point. This is true, for example, of $\mathbb{S}^2 \subseteq \mathbb{R}^3$. However, every compact Lie group has a global orthonormal frame (this follows from the comment after Equation (5.5) below).

The definitions above have been formulated for general embedded manifolds, but in the setting of Lie groups, one can normally restrict attention to what happens at the identity $e \in G$ and get the rest via translation within the group. Specifically, any vector $X \in T_e(G)$ defines a vector field $\widetilde{X}$ on G as follows. For $g \in G$ fixed, let $L_g : G \to G$ denote the map given by $L_g(h) = gh$. Then for any $h \in G$, define

$$\widetilde{X}(h) := (L_{h*})_e X = \frac{d}{dt}[h\gamma(t)]\bigg|_{t=0},$$

for any curve γ in G with $\gamma(0) = e$ and $\gamma'(0) = X$. The vector field $\widetilde{X}$ acts as a differential operator by

$$\widetilde{X}(f)(h) = \frac{d}{dt} f(h\gamma(t))\bigg|_{t=0},$$

since $\gamma_h(t) = h\gamma(t)$ is a curve with $\gamma_h(0) = h$ and

$$\gamma_h'(0) = \frac{d}{dt}[h\gamma(t)]\bigg|_{t=0} = \widetilde{X}(h).$$

A vector field Y on G with the property that for any $g \in G, L_{g*}Y = Y$ is called a **(left) invariant vector field**. For any $X \in T_e(G)$, the extension $\widetilde{X}$ described above gives an invariant vector field on G, since

$$[L_{g*}\widetilde{X}](gh) = (L_{g*})_h(\widetilde{X}(h)) = (L_{g*})_h((L_{h*})_e X) = (L_{gh*})_e X.$$

Conversely, given an invariant vector field Y, $Y(h) = (L_{h*})_e(Y(e)) = \widetilde{Y(e)}(h)$, and so the mapping $X \mapsto \widetilde{X}$ gives a bijection between invariant vector fields and elements of $T_e(G)$; either of these vector spaces may be referred to as the Lie algebra of G.

From an intrinsic differential geometric point of view, this is a fine definition of $\widetilde{X}$, but because our Riemannian metric is the one inherited from the ambient Euclidean space, it helps to also have a more concrete perspective on $\widetilde{X}$. As above, let $\gamma : (-\epsilon, \epsilon) \to G$ be a curve with $\gamma(0) = e$ and $\gamma'(0) = X$, and for each $h \in G$, define γ_h by $\gamma_h(t) = h\gamma(t)$. Then using the Euclidean structure gives that

$$\widetilde{X}(h) = \gamma_h'(0) = \lim_{s \to 0} \frac{\gamma_h(s) - h}{s} = h\left(\lim_{s \to 0} \frac{\gamma(s) - e}{s}\right) = h\gamma'(0) = hX.$$

In particular, this means that if G is one of the classical compact groups so that the inner product on $T_I(G)$ is the real part of the Hilbert–Schmidt inner product, then for any $h \in G$ and any $\widetilde{X}$ and $\widetilde{Y}$ defined as above,

$$g_h(\widetilde{X}(h), \widetilde{Y}(h)) = \mathrm{Re}(\mathrm{Tr}(hXY^*h^*)) = \mathrm{Re}(\mathrm{Tr}(XY^*)) = \langle X, Y \rangle . \tag{5.5}$$

In particular, if X and Y are orthogonal elements of $T_I(G)$, then $\widetilde{X}$ and $\widetilde{Y}$ are orthogonal at every point of G.

Given two vector fields X and Y on M, there is a unique vector field $[X, Y]$, called the **Lie Bracket** of X and Y, such that

$$[X, Y](f) = X(Y(f)) - Y(X(f)).$$

The fact that this is a vector field is not obvious, and is in fact a bit surprising, since vector fields can be thought of as first order differential operators, but this looks like a second-order operator. Indeed, just XY and YX by themselves are not vector fields, but in the case of the Lie bracket, the second-order parts cancel out.

Exercise 5.10 Show that for $F : M \to N$ a smooth map between manifolds and X and Y vector fields on M,

$$[F_*X, F_*Y] = F_*[X, Y].$$

It follows in particular from the exercise that on a Lie group G, if X, Y are invariant vector fields, then so is $[X, Y]$. Since for given $X, Y \in T_e(G)$, $\widetilde{X}$ and $\widetilde{Y}$ are invariant, this means that there must be some vector $Z \in T_e(G)$ such that $\widetilde{Z} = [\widetilde{X}, \widetilde{Y}]$. The identity of this vector Z is given in the following lemma.

Lemma 5.11 *Let G be one of the classical compact groups and $\mathfrak{g} = T_e(G)$ its Lie algebra. Let $X, Y \in \mathfrak{g}$, and define*

$$[X, Y] = XY - YX,$$

where here XY refers to the matrix product of X and Y. Then $[X, Y] \in \mathfrak{g}$ and

$$\widetilde{[X, Y]} = [\widetilde{X}, \widetilde{Y}],$$

where the expression on the right is the Lie bracket of the vector fields $\widetilde{X}$ and $\widetilde{Y}$ as defined above.

Proof We will verify that $[X, Y] \in \mathfrak{g}$ in the case of $G = \mathbb{SU}(n)$; the remaining cases are essentially the same. Recall that

$$\mathfrak{su}(n) = \left\{X \in \mathrm{M}_n(\mathbb{C}) : X + X^* = 0, \mathrm{Tr}(X) = 0\right\}.$$

Given $X, Y \in \mathfrak{su}(n)$,

$$\begin{aligned}[X, Y] + [X, Y]^* &= XY - YX + Y^*X^* - X^*Y^* \\ &= XY - YX + YX - XY = 0,\end{aligned}$$

and

$$\mathrm{Tr}([X, Y]) = \mathrm{Tr}(XY - YX) = 0.$$

To verify the claim that $\widetilde{[X, Y]} = [\widetilde{X}, \widetilde{Y}]$, fix a smooth function f on G and an element $g \in G$. Then $f(g(I + Z))$ is a smooth function of Z, and so by Taylor's theorem, we can write

$$f(g(I + Z)) = c_0 + c_1(Z) + B(Z, Z) + R(Z),$$

where c_1 is a linear function, B is a symmetric bilinear form, $|R(X)| \leq c_3\|X\|_{H.S.}^3$ for some $c_3 > 0$, and c_0, c_1, B, and R all depend only on f and g. Expanding the exponential gives that

$$f(g\exp(Z)) = c_0 + c_1(Z) + \widetilde{B}(Z, Z) + \widetilde{R}(Z),$$

for another symmetric bilinear $\widetilde{B}$ and $\widetilde{R}$ which vanishes to third order. Then for $Z \in T_I(G)$,

$$\begin{aligned}\widetilde{Z}(f)(g) &= \frac{d}{dt} f(g\exp(tZ))\bigg|_{t=0} \\ &= \frac{d}{dt}\left(c_0 + c_1(tZ) + \widetilde{B}(tZ, tZ) + \widetilde{R}(tZ)\right)\bigg|_{t=0} = c_1(Z).\end{aligned}$$

Now,

$$\begin{aligned}
\widetilde{X}(\widetilde{Y}(f))(g) &= \frac{d}{dt}\widetilde{Y}(f)(g\exp(tX))\Big|_{t=0} \\
&= \frac{d}{dt}\frac{d}{ds}f(g\exp(tX)\exp(sY))\Big|_{s=0}\Big|_{t=0} \\
&= \frac{d}{dt}\frac{d}{ds}f(g(I+tX+r(tX))(I+sY+r(sY))\Big|_{s=0}\Big|_{t=0},
\end{aligned}$$

where $\|r(Z)\|_{H.S.} \leq c_2\|Z\|_{H.S.}^2$ for some $c_2 > 0$ and $Z \in G$. Proceeding as before, and writing only the terms of the expansion that give some contribution in the limit,

$$\begin{aligned}
\widetilde{X}(\widetilde{Y}(f))(g) &= \frac{d}{dt}\frac{d}{ds}(c_0 + c_1(tX+sY+stXY) + B(tX+sY, tX+sY))\Big|_{s=0}\Big|_{t=0} \\
&= c_1(XY) + 2B(X,Y).
\end{aligned}$$

It follows that

$$[\widetilde{X},\widetilde{Y}](f)(g) = c_1(XY - YX) = \widetilde{[X,Y]}(f)(g),$$

which proves the claim. □

We still need a few more notions in order to get to curvature. First, a **connection** ∇ on M is a way of differentiating one vector field in the direction of another: a connection ∇ is a bilinear form on vector fields that assigns to vector fields X and Y a new vector field $\nabla_X Y$, such that for any smooth function $f : M \to \mathbb{R}$,

$$\nabla_{fX} Y = f\nabla_X Y \qquad \text{and} \qquad \nabla_X(fY) = f\nabla_X(Y) + X(f)Y.$$

A connection is called **torsion-free** if

$$\nabla_X Y - \nabla_Y X = [X,Y]. \tag{5.6}$$

There is a special connection on a Riemannian manifold, called the **Levi-Civita connection**, which is the unique torsion-free connection with the property that

$$X(g(Y,Z)) = g(\nabla_X Y, Z) + g(Y, \nabla_X Z). \tag{5.7}$$

This property may look not obviously interesting, but geometrically, it is a compatibility condition of the connection ∇ with g. There is a notion of transporting a vector field in a "parallel way" along a curve, which is defined by the connection. The condition above means that the inner product defined by g of two vector fields at a point is unchanged if you parallel-transport the vector fields (using ∇ to define "parallel") along any curve.

Finally, we can define the **Riemannian curvature tensor** $R(X, Y)$: to each pair of vector fields X and Y on M, we associate an operator $R(X, Y)$ on vector fields defined by

$$R(X, Y)(Z) := \nabla_X(\nabla_Y Z) - \nabla_Y(\nabla_X Z) - \nabla_{[X,Y]}Z.$$

The **Ricci curvature tensor** is the function $\mathrm{Ric}(X, Y)$ on M which, at each point $p \in M$, is the trace of the linear map on T_pM defined by $Z \mapsto R(Z, Y)(X)$. In orthonormal local coordinates $\{L_i\}$,

$$\mathrm{Ric}(X, Y) = \sum_i g(R(X, L_i)L_i, Y).$$

(Seeing that this coordinate expression is right involves using some of the symmetries of R.)

The Bakry–Émery Criterion

The Bakry–Émery criterion can be made more general, but for our purposes it suffices to formulate it as follows.

Theorem 5.12 (The Bakry–Émery curvature criterion) *Let (M, g) be a compact, connected, m-dimensional Riemannian manifold with normalized volume measure μ. Suppose that there is a constant $c > 0$ such that for each $p \in M$ and each $v \in T_pM$,*

$$\mathrm{Ric}_p(v, v) \geq \frac{1}{c} g_p(v, v).$$

Then μ satisfies a log-Sobolev inequality with constant c.

The following proposition together with the Bakry–Émery criterion leads to log-Sobolev inequalities, and thus concentration of measure, on most of the classical compact groups.

Proposition 5.13 *If G_n is one of $\mathbb{SO}(n)$, $\mathbb{SO}^-(n)$ $\mathbb{SU}(n)$, or $\mathbb{Sp}(2n)$, then for each $U \in G_n$ and each $X \in T_U G_n$,*

$$\mathrm{Ric}_U(X, X) = c_{G_n} g_U(X, X),$$

where g_U is the Hilbert–Schmidt metric and c_{G_n} is given by

G	c_G
$\mathbb{SO}(n), \mathbb{SO}^-(n)$	$\frac{n-2}{4}$
$\mathbb{SU}(n)$	$\frac{n}{2}$
$\mathbb{Sp}(2n)$	$n+1$

For the curvature computation, it is simplest to work with the symplectic group in its quaternionic form, with the Lie algebra

$$\mathfrak{su}_{\mathbb{H}}(n) = \left\{X \in \mathrm{M}_n(\mathbb{H}) : X + X^* = 0\right\},$$

where

$$\mathbb{H} = \{a + b\mathbf{i} + c\mathbf{j} + d\mathbf{k} : a, b, c, d \in \mathbb{R}\}$$

is the skew field of quaternions, $\overline{(a + b\mathbf{i} + c\mathbf{j} + d\mathbf{k})} = a - b\mathbf{i} - c\mathbf{j} - d\mathbf{k}$, and the (real) inner product on $\mathfrak{su}_{\mathbb{H}}(n)$ is given by $\langle X, Y\rangle = \mathrm{Tr}(XY^*)$.

The following proposition is a key part of the proof of Proposition 5.13.

Proposition 5.14 *Let $X \in \mathfrak{g}$, where $\mathfrak{g}$ is one of $\mathfrak{so}(n)$, $\mathfrak{su}(n)$, or $\mathfrak{su}_{\mathbb{H}}(n)$, and let $\{L_\alpha\}_{\alpha\in A}$ be an orthonormal basis of $\mathfrak{g}$. Then*

$$-\frac{1}{4}\sum_{\alpha\in A}[[X, L_\alpha], L_\alpha] = \left(\frac{\beta(n+2)}{4} - 1\right)X,$$

where $\beta = 1, 2, 4$ as $\mathfrak{g}$ is $\mathfrak{so}(n)$, $\mathfrak{su}(n)$, or $\mathfrak{su}_{\mathbb{H}}(n)$.

Proof We first observe that the expression on the left is independent of the choice of orthonormal basis. Indeed, each $\mathfrak{g}$ is a real inner product space (with the inner product given by $\langle X, Y\rangle = \mathrm{Re}(\mathrm{Tr}(XY^*))$). If $\{K_\alpha\}_{\alpha\in A}$ is a second orthonormal basis of $\mathfrak{g}$, then there is an orthogonal matrix $U = [u_{\alpha,\beta}]_{\alpha,\beta\in A}$ such that

$$K_\beta = \sum_\alpha u_{\alpha,\beta} L_\alpha.$$

Then

$$\sum_{\beta\in A}[[X, K_\beta], K_\beta] = \sum_{\beta\in A}\sum_{\alpha_1,\alpha_2\in A} u_{\alpha_1,\beta}u_{\alpha_2,\beta}[[X, L_{\alpha_1}], L_{\alpha_2}] = \sum_{\alpha\in A}[[X, L_\alpha], L_\alpha]$$

by the orthogonality of the matrix U.

Note that $\mathfrak{so}(1) = \mathfrak{su}(1) = \{0\}$. It is easy to check the claim for $\mathfrak{su}_{\mathbb{H}}(1)$ with the basis $\{\mathbf{i}, \mathbf{j}, \mathbf{k}\}$, so in what follows, we will assume $n \geq 2$.

If $1 \leq j, k \leq n$, we use E_{jk} to denote the matrix with 1 in the j-k entry and zeros otherwise. For $q \in \{\mathbf{i}, \mathbf{j}, \mathbf{k}\}$ and $1 \leq \ell < n$, define

$$D_\ell^q := \frac{q}{\sqrt{\ell + \ell^2}} \left(\sum_{r=1}^{\ell} E_{rr} - \ell E_{\ell+1,\ell+1} \right)$$

and let $D_n^q := \frac{q}{\sqrt{n}} I_n$. Define $D_\ell := D_\ell^{\mathbf{i}}$.

For $q \in \{1, \mathbf{i}, \mathbf{j}, \mathbf{k}\}$ and $1 \leq \ell < r \leq n$, define

$$F_{\ell r}^q := \frac{q}{\sqrt{2}} E_{\ell,r} - \frac{\overline{q}}{\sqrt{2}} E_{r,\ell}.$$

Let $F_{\ell r} := F_{\ell r}^1$ and let $G_{\ell r} = F_{\ell r}^{\mathbf{i}}$. Then

- $\{F_{\ell r} : 1 \leq \ell < r \leq n\}$ is an orthonormal basis of $\mathfrak{so}(n)$;
- $\{D_\ell : 1 \leq \ell < n\} \cup \{F_{\ell r}, G_{\ell r} : 1 \leq \ell < r \leq n\}$ is an orthonormal basis of $\mathfrak{su}(n)$;
- $\left\{D_\ell^q : 1 \leq \ell \leq n, q \in \{\mathbf{i}, \mathbf{j}, \mathbf{k}\}\right\} \cup \left\{F_{\ell r}^q : 1 \leq \ell < r \leq n, q \in \{1, \mathbf{i}, \mathbf{j}, \mathbf{k}\}\right\}$ is an orthonormal basis of $\mathfrak{su}_{\mathbb{H}}(n)$.

It suffices to verify the claim for these orthonormal bases $\{L_\alpha\}_{\alpha \in A}$. We can make a further simplification as follows: suppose that the claimed formula holds for a particular orthonormal basis $\{L_\alpha\}_{\alpha \in A}$ and a particular choice of X. Let $U \in G$. Then

$$\begin{aligned} \left(\frac{\beta(n+2)}{4} - 1 \right) UXU^* &= -\frac{1}{4} \sum_{\alpha \in A} U[X, L_\alpha], L_\alpha] U^* \\ &= -\frac{1}{4} \sum_{\alpha \in A} [[UXU^*, UL_\alpha U^*], UL_\alpha U^*]. \end{aligned}$$

It is easy to check that $\{UL_\alpha U^*\}_{\alpha \in A}$ is again an orthonormal basis of $\mathfrak{g}$, and so we have that if the claimed formula holds for X, then it holds for UXU^*.

Take $X = F_{12}$. We will show that the collection $\{UXU^* : U \in G\}$ spans $\mathfrak{g}$, so that it finally suffices to verify the claimed formula for the orthonormal bases listed above and the single element $X = F_{12}$.

All of the $F_{\ell r}$ are of the form UXU^* for a permutation matrix U. Choosing

$$U = \frac{1}{\sqrt{n}} \begin{bmatrix} 1+q & 0 \\ 0 & 1-q \end{bmatrix} \oplus I_{n-2}$$

for $q \in \{\mathbf{i}, \mathbf{j}, \mathbf{k}\}$ gives

$$UXU^* = \frac{1}{n} F_{12}^q,$$

and further conjugation by permutation matrices yields (multiples of) all the $F_{\ell r}^1$. Choosing

$$U = \frac{1}{\sqrt{n+2}} \begin{bmatrix} q & 1 \\ 1 & q \end{bmatrix} \oplus I_{n-2}$$

for $q \in \{\mathbf{i}, \mathbf{j}, \mathbf{k}\}$ gives

$$UXU^* = \left(\frac{2}{n+2}\right) D_1^q;$$

further conjugation by permutation matrices yields all matrices with one 1 and one -1 on the diagonal. By taking linear combinations, this yields all of the D_ℓ^q for $1 \le \ell < n$. Finally, note that

$$\begin{bmatrix} 1 & 0 \\ 0 & \mathbf{j} \end{bmatrix} \begin{bmatrix} \mathbf{i} & 0 \\ 0 & -\mathbf{i} \end{bmatrix} \begin{bmatrix} 1 & 0 \\ 0 & -\mathbf{j} \end{bmatrix} = \begin{bmatrix} \mathbf{i} & 0 \\ 0 & \mathbf{i} \end{bmatrix};$$

taking $U = \begin{bmatrix} 1 & 0 \\ 0 & \mathbf{j} \end{bmatrix} \oplus I_{n-2}$, and then taking linear combinations of conjugations of UXU^* by permutation matrices results in a (real) multiple of $D_n^{\mathbf{i}}$; the remaining D_n^q can be obtained similarly.

All that remains is to finally verify the formula for $X = F_{12}$. Now, F_{12} commutes with all the D_ℓ^q with $\ell > 1$ and all the $F_{\ell r}^q$ with $2 < \ell < r \le n$. For $q \in \{\mathbf{i}, \mathbf{j}, \mathbf{k}\}$,

$$[[F_{12}, F_{12}^q], F_{12}^q] = [[F_{12}, D_1^q], D_1^q] = -2F_{12}.$$

If $1 \le \ell \le 2 < r \le n$, then

$$[[F_{12}, F_{\ell r}^q], F_{\ell r}^q] = -\frac{1}{2} F_{12}.$$

From this it is clear that $\sum_\alpha [[X, L_\alpha], L_\alpha]$ is some multiple of X; collecting terms yields exactly the claimed constant. □

We now give the proof of Proposition 5.13.

Proof of Proposition 5.13 First we observe that the defining properties (5.6) and (5.7) of the Levi-Civita connection imply that for all vector fields X, Y, Z,

$$\begin{aligned} 2g(\nabla_X Y, Z) &= X(g(Y,Z)) + Y(g(Z,X)) - Z(g(X,Y)) \\ &\quad + g([X,Y],Z) + g([Z,X],Y) + g(X,[Z,Y]). \end{aligned}$$

In particular, since $g(\widetilde{X}, \widetilde{Y})$ is constant for any vectors $X, Y \in T_I(G)$, it follows that

$$\begin{aligned}
2g(\nabla_{\widetilde{X}}\widetilde{Y}, \widetilde{Z}) &= g([\widetilde{X}, \widetilde{Y}], \widetilde{Z}) + g([\widetilde{Z}, \widetilde{X}], \widetilde{Y}) + g(\widetilde{X}, [\widetilde{Z}, \widetilde{Y}]) \\
&= \langle [X, Y], Z\rangle + \langle [Z, X], Y\rangle + \langle X, [Z, Y]\rangle
\end{aligned}$$

for all $X, Y, Z \in T_I(G)$. Using the fact that $X, Y, Z \in T_I(G)$ so that, e.g., $X^* = -X$ leads to the further simplification

$$2g(\nabla_{\widetilde{X}}\widetilde{Y}, \widetilde{Z}) = \langle [X, Y], Z\rangle = g([\widetilde{X}, \widetilde{Y}], \widetilde{Z}).$$

Taking $Z = L_\alpha$ for $\{L_\alpha\}_{\alpha \in A}$ an orthonormal basis of $T_I(G)$ and summing over α gives that

$$\nabla_{\widetilde{X}}\widetilde{Y} = \frac{1}{2}\widetilde{[X, Y]}.$$

Then

$$\begin{aligned}
R(\widetilde{X}, \widetilde{L_\alpha})\widetilde{L_\alpha} &= \nabla_{\widetilde{X}}(\nabla_{\widetilde{L_\alpha}}\widetilde{L_\alpha}) - \nabla_{\widetilde{L_\alpha}}(\nabla_{\widetilde{X}}\widetilde{L_\alpha}) - \nabla_{[\widetilde{X}, \widetilde{L_\alpha}]}\widetilde{L_\alpha} \\
&= -\frac{1}{4}\big[[X, L_\alpha], L_\alpha\big].
\end{aligned}$$

The coordinate expression for the Ricci curvature together with Proposition 5.14 now gives that

$$\begin{aligned}
\mathrm{Ric}(\widetilde{X}, \widetilde{X}) &= -\frac{1}{4}\sum_{\alpha \in A} \langle [[X, L_\alpha], X_\alpha], X\rangle \\
&= \left(\frac{\beta(n+2)}{4} - 1\right)\langle X, X\rangle = \left(\frac{\beta(n+2)}{4} - 1\right) g(\widetilde{X}, \widetilde{X}).
\end{aligned}$$

□

Log-Sobolev inequalities, and hence concentration inequalities, now follow immediately from the Bakry–Émery Theorem for the groups listed above; i.e., all of the classical compact groups except $\mathbb{O}(n)$ and $\mathbb{U}(n)$. On $\mathbb{O}(n)$, we cannot expect more and indeed more is not true, because $\mathbb{O}(n)$ is disconnected. We do have the best that can be hoped for, namely concentration on each of the pieces. In the case of $\mathbb{U}(n)$, though, there is the same kind of concentration that we have on $\mathbb{SU}(n)$. There is no nonzero lower bound on the Ricci curvature on $\mathbb{U}(n)$: $\mathrm{Ric}(\widetilde{X}, \widetilde{X}) = 0$ when $X = iI \in T_I(\mathbb{U}(n))$. Instead, one can obtain a log-Sobolev inequality on $\mathbb{U}(n)$ from the one on $\mathbb{SU}(n)$ via a coupling argument. The following slightly nonstandard coupling of the Haar measures on $\mathbb{SU}(n)$ and $\mathbb{U}(n)$ is the key to obtaining the right dimensional dependence in the constant.

Lemma 5.15 *Let θ be uniformly distributed in $\left[0, \frac{2\pi}{n}\right]$ and let $V \in \mathbb{SU}(n)$ be uniformly distributed, with θ and V independent. Then $e^{i\theta}V$ is uniformly distributed in $\mathbb{U}(n)$.*

Proof Let X be uniformly distributed in $[0, 1)$, K uniformly distributed in $\{0, \ldots, n-1\}$, and V uniformly distributed in $\mathbb{SU}(n)$ with (X, K, V) independent. Consider

$$U = e^{2\pi iX/n} e^{2\pi iK/n} V.$$

On one hand, it is easy to see that $(X + K)$ is uniformly distributed in $[0, n]$, so that $e^{2\pi i(X+K)/n}$ is uniformly distributed on $\mathbb{S}^1$. Thus $U \overset{d}{=} \omega V$ for ω uniform in $\mathbb{S}^1$ and independent of V. It is clear that the distribution of ωV is translation-invariant on $\mathbb{U}(n)$, so that U is Haar-distributed.

On the other hand, if I_n is the $n \times n$ identity matrix, then $e^{2\pi iK/n} I_n \in \mathbb{SU}(n)$. By the translation invariance of Haar measure on $\mathbb{SU}(n)$ this implies that $e^{2\pi iK/n} V \overset{d}{=} V$, and so $e^{2\pi iX/n} V \overset{d}{=} U$. □

The log-Sobolev inequality on $\mathbb{U}(n)$ now follows using this coupling together with the tensorization property of LSI, as follows.

Proof of LSI on $\mathbb{U}(n)$ First, for the interval $[0, 2\pi]$ equipped with its standard metric and uniform measure, the optimal constant in (5.2) for functions f with $f(0) = f(2\pi)$ is known to be 1; see, e.g., [104]. This fact completes the proof in the case $n = 1$; from now on, assume that $n \geq 2$.

Suppose that $f : [0, \pi] \to \mathbb{R}$ is locally Lipschitz, and define a function $\tilde{f} : [0, 2\pi] \to \mathbb{R}$ by reflection:

$$\tilde{f}(x) := \begin{cases} f(x), & 0 \leq x \leq \pi; \\ f(2\pi - x), & \pi \leq x \leq 2\pi. \end{cases}$$

Then $\tilde{f}$ is locally Lipschitz and $\tilde{f}(2\pi) = \tilde{f}(0)$, so $\tilde{f}$ satisfies a LSI for uniform measure on $[0, 2\pi]$ with constant 1. If $\mu_{[a,b]}$ denotes uniform (probability) measure on $[a, b]$, then

$$\operatorname{Ent}_{\mu_{[0,2\pi]}}(\tilde{f}^2) = \operatorname{Ent}_{\mu_{[0,\pi]}}(f^2),$$

and

$$\frac{1}{2\pi} \int_0^{2\pi} |\nabla \tilde{f}(x)|^2 dx = \frac{1}{\pi} \int_0^{\pi} |\nabla f(x)|^2 dx,$$

so f itself satisfies a LSI for uniform measure on $[0, \pi]$ with constant 1 as well.

It then follows by a scaling argument that the optimal logarithmic Sobolev constant on $\left[0, \frac{\pi\sqrt{2}}{\sqrt{n}}\right)$ is $2/n$ (for $g : \left[0, \frac{\pi\sqrt{2}}{\sqrt{n}}\right) \to \mathbb{R}$, apply the LSI to $g\left(\sqrt{\frac{2}{n}}x\right)$ and rearrange it to get the LSI on $\left[0, \frac{\pi\sqrt{2}}{\sqrt{n}}\right)$).

Combining Proposition 5.13 with the Bakry–Émery criterion shows that $\mathbb{SU}(n)$ satisfies a log-Sobolev inequality with constant $2/n$ when equipped with its geodesic distance, and hence also when equipped with the Hilbert–Schmidt metric. By the tensorization property of log-Sobolev inequalities in product spaces (Lemma 5.7), the product space $\left[0, \frac{\pi\sqrt{2}}{\sqrt{n}}\right) \times \mathbb{SU}(n)$, equipped with the L_2-sum metric, satisfies a log-Sobolev inequality with constant $2/n$ as well.

Define the map $F : \left[0, \frac{\pi\sqrt{2}}{\sqrt{n}}\right) \times \mathbb{SU}(n) \to \mathbb{U}(n)$ by $F(t, V) = e^{\sqrt{2}it/\sqrt{n}}V$. By Lemma 5.15, the push-forward via F of the product of uniform measure on $\left[0, \frac{\pi\sqrt{2}}{\sqrt{n}}\right)$ with uniform measure on $\mathbb{SU}(n)$ is uniform measure on $\mathbb{U}(n)$. Moreover, this map is $\sqrt{3}$-Lipschitz:

$$\begin{aligned}
\left\| e^{\sqrt{2}it_1/\sqrt{n}}V_1 - e^{\sqrt{2}it_2/\sqrt{n}}V_2 \right\|_{HS} &\leq \left\| e^{\sqrt{2}it_1/\sqrt{n}}V_1 - e^{\sqrt{2}it_1/\sqrt{n}}V_2 \right\|_{HS} \\
&\quad + \left\| e^{\sqrt{2}it_1/\sqrt{n}}V_2 - e^{\sqrt{2}it_2/\sqrt{n}}V_2 \right\|_{HS} \\
&= \|V_1 - V_2\|_{HS} + \left\| e^{\sqrt{2}it_1/\sqrt{n}}I_n - e^{\sqrt{2}it_2/\sqrt{n}}I_n \right\|_{HS} \\
&\leq \|V_1 - V_2\|_{HS} + \sqrt{2}\,|t_1 - t_2| \\
&\leq \sqrt{3}\sqrt{\|V_1 - V_2\|_{HS}^2 + |t_1 - t_2|^2}.
\end{aligned}$$

It now follows from Lemma 5.6 that Haar measure on $\mathbb{U}(n)$ satisfies a logarithmic Sobolev inequality with constant $(\sqrt{3})^2\frac{2}{n} = \frac{6}{n}$. □

Summarizing, we have the following.

Theorem 5.16 *The matrix groups and cosets $\mathbb{SO}(n)$, $\mathbb{SO}^-(n)$, $\mathbb{SU}(n)$, $\mathbb{U}(n)$, and $\mathbb{Sp}(2n)$ with Haar probability measure and the Hilbert–Schmidt metric, satisfy logarithmic Sobolev inequalities with the following constants:*

G	C_G
$\mathbb{SO}(n)$, $\mathbb{SO}^-(n)$	$\frac{4}{n-2}$
$\mathbb{SU}(n)$	$\frac{2}{n}$
$\mathbb{U}(n)$	$\frac{6}{n}$
$\mathbb{Sp}(2n)$	$\frac{1}{n+1}$

Recall from Lemma 1.3 that the the geodesic distance on $\mathbb{U}(n)$ is bounded above by $\pi/2$ times the Hilbert–Schmidt distance. Thus Theorem 5.16 implies, for example, that $\mathbb{U}(n)$ equipped with the geodesic distance also satisfies a log-Sobolev inequality, with constant $3\pi^2/2n$.

The following summarizes the concentration properties of Haar measure on the classical compact groups that follow from the log-Sobolev constants above together with Theorem 5.9.

Theorem 5.17 *Given $n_1, \ldots, n_k \in \mathbb{N}$, let $X = G_{n_1} \times \cdots \times G_{n_k}$, where for each of the n_i, G_{n_i} is one of $\mathbb{SO}(n_i)$, $\mathbb{SO}^-(n_i)$, $\mathbb{SU}(n_i)$, $\mathbb{U}(n_i)$, or $\mathbb{Sp}(2n_i)$. Let X be equipped with the L^2-sum of Hilbert–Schmidt metrics on the G_{n_i}. Suppose that $F : X \to \mathbb{R}$ is L-Lipschitz, and that $\{U_j \in G_{n_j} : 1 \leq j \leq k\}$ are independent, Haar-distributed random matrices. Then for each $t > 0$,*

$$\mathbb{P}\big[F(U_1, \ldots, U_k) \geq \mathbb{E}F(U_1, \ldots, U_k) + t\big] \leq e^{-(n-2)t^2/24L^2},$$

where $n = \min\{n_1, \ldots, n_k\}$.

5.4 Concentration of the Spectral Measure

The following theorem quantifies the rate of convergence of the empirical spectral measure of a random unitary matrix to the uniform measure on the circle, and more generally the empirical spectral measure of a power of a random unitary matrix. Recall that W_p denotes the L_p Kantorovich distance between measures (see Section 2.1).

Theorem 5.18 *Let $\mu_{n,m}$ be the spectral measure of U^m, where $1 \leq m \leq n$ and $U \in \mathbb{U}(n)$ is distributed according to Haar measure, and let ν denote the uniform measure on $\mathbb{S}^1$. Then for each $p \geq 1$,*

$$\mathbb{E}W_p(\mu_{n,m}, \nu) \leq Cp\frac{\sqrt{m\left[\log\left(\frac{n}{m}\right) + 1\right]}}{n},$$

where $C > 0$ is an absolute constant.

For each $t > 0$,

$$\mathbb{P}\left[W_p(\mu_{n,m}, \nu) \geq C\frac{\sqrt{m\left[\log\left(\frac{n}{m}\right) + 1\right]}}{n} + t\right] \leq \exp\left[-\frac{n^2t^2}{24m}\right]$$

for $1 \leq p \leq 2$ and

$$\mathbb{P}\left[W_p(\mu_{n,m}, \nu) \geq Cp\frac{\sqrt{m\left[\log\left(\frac{n}{m}\right) + 1\right]}}{n} + t\right] \leq \exp\left[-\frac{n^{1+2/p}t^2}{24m}\right]$$

for $p > 2$, where $C > 0$ is an absolute constant.

The change in behavior observed above at $p = 2$ is typical for the Kantorovich distances.

By a simple application of the Borel-Cantelli lemma, one gets an almost sure rate of convergence, as follows.

Corollary 5.19 *Suppose that for each n, $U_n \in \mathbb{U}(n)$ is Haar-distributed and $1 \leq m_n \leq n$. Let ν denote the uniform measure on $\mathbb{S}^1$. There is an absolute constant C such that given $p \geq 1$, with probability 1, for all sufficiently large n,*

$$W_p(\mu_{n,m_n}, \nu) \leq C\frac{\sqrt{m_n \log(n)}}{n}$$

if $1 \leq p \leq 2$ and

$$W_p(\mu_{n,m_n}, \nu) \leq Cp\frac{\sqrt{m_n \log(n)}}{n^{\frac{1}{2}+\frac{1}{p}}}$$

if $p > 2$.

Observe in particular the change in behavior of the bound as m grows: for $m = 1$,

$$W_p(\mu_n, \nu) \leq \frac{C\sqrt{\log(n)}}{n}.$$

Since μ_n is supported on n points, this estimate means that the eigenvalues are very regularly spaced; $W_p(\mu_n, \nu)$ is only logarithmically larger than the distance from ν to a discrete measure on n points exactly evenly spaced around the circle (which is exactly $\frac{\pi}{n}$).

At the opposite extreme, when $m = n$ the bound becomes

$$W_p(\mu_{n,n}, \nu) \leq \frac{C}{\sqrt{n}}.$$

This result is in fact classical (and known to be sharp), since by Theorem 3.14, $\mu_{n,n}$ is exactly the empirical measure of n i.i.d. uniform random points on the circle. One would thus expect the eigenvalues of U^n to be considerably less regular than those of U, and indeed this and the intermediate phenomena can be observed in the simulation shown in Figure 5.1.

The first step in proving Theorem 5.18 is to prove a concentration inequality for the number $\mathcal{N}_\theta^{(m)}$ of eigenangles of U^m in $[0, \theta)$. Such a concentration result is an easy consequence of Theorems 4.1 and 3.14. Specifically, recall that since the eigenangles of a random unitary matrix are a determinantal projection process, it follows from Theorem 4.1 that

$$\mathcal{N}_\theta^{(1)} \stackrel{d}{=} \sum_{k=1}^{n} \xi_k,$$

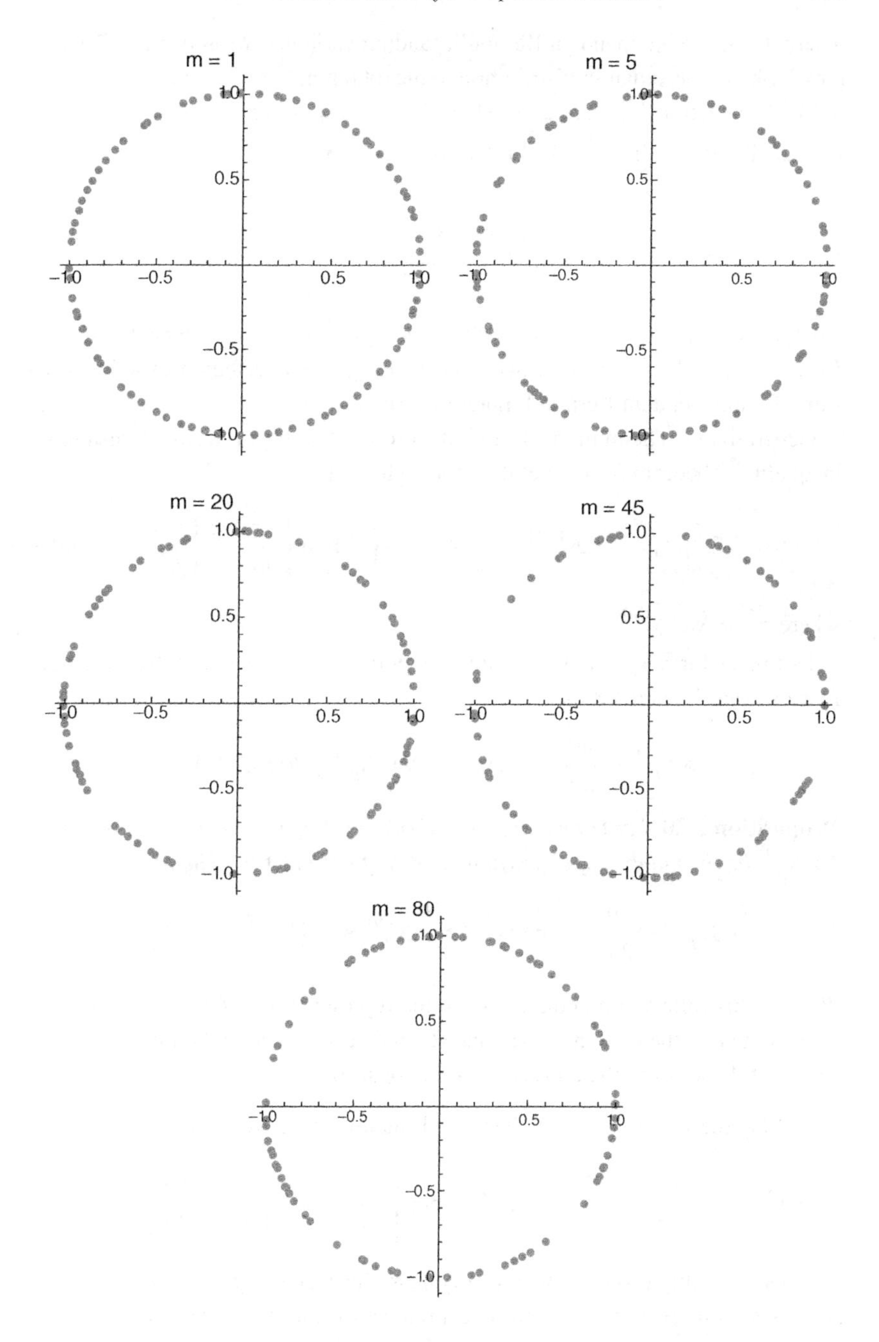

Figure 5.1 The eigenvalues of U^m for U an 80×80 random unitary matrix.

where the ξ_k are independent Bernoulli random variables. Moreover, by Theorem 3.14, $\mathcal{N}_\theta^{(m)}$ is equal in distribution to the total number of eigenvalue angles in $[0,\theta)$ of each of $U_0 \dots, U_{m-1}$, where $U_0, \dots, U_{m-1}$ are independent and U_j is Haar-distributed in $\mathbb{U}\left(\left\lceil \frac{n-j}{m} \right\rceil\right)$; that is,

$$\mathcal{N}_\theta^{(m)} \stackrel{d}{=} \sum_{j=0}^{m-1} \mathcal{N}_{j,\theta}, \tag{5.8}$$

where the $\mathcal{N}_{j,\theta}$ are the independent counting functions corresponding to $U_0, \dots, U_{m-1}$. It is therefore also true that $\mathcal{N}_\theta^{(m)}$ is distributed exactly as a sum of n independent Bernoulli random variables.

Generalizing Theorem 4.11 and its proof, it follows from Bernstein's inequality (Theorem 5.1) to get that, for each $t > 0$,

$$\mathbb{P}\left[\left|\mathcal{N}_\theta^{(m)} - \mathbb{E}\mathcal{N}_\theta^{(m)}\right| > t\right] \leq 2\exp\left(-\min\left\{\frac{t^2}{4\sigma^2}, \frac{t}{2}\right\}\right), \tag{5.9}$$

where $\sigma^2 = \operatorname{Var} \mathcal{N}_\theta^{(m)}$.

Estimates for $\mathbb{E}\mathcal{N}_\theta^{(m)}$ and σ^2 follow easily from the $m = 1$ case. Recall from Propositions 4.7 and 4.8 that

$$\mathbb{E}\mathcal{N}_\theta^{(1)} = \frac{n\theta}{2\pi} \qquad \text{and} \qquad \operatorname{Var} \mathcal{N}_\theta^{(1)} \leq \log(n) + 1.$$

Proposition 5.20 *Let U be uniform in $\mathbb{U}(n)$ and $1 \leq m \leq n$. For $\theta \in [0, 2\pi)$, let $\mathcal{N}_\theta^{(m)}$ be the number of eigenvalue angles of U^m in $[0,\theta)$. Then*

$$\mathbb{E}\mathcal{N}_\theta^{(m)} = \frac{n\theta}{2\pi} \qquad \textit{and} \qquad \operatorname{Var} \mathcal{N}_\theta^{(m)} \leq m\left(\log\left(\frac{n}{m}\right) + 1\right).$$

Proof This follows immediately from the representation of $\mathcal{N}_\theta^{(m)}$ in equation (5.8); note that the n/m in the variance bound, as opposed to the more obvious $\lceil n/m \rceil$, follows from the concavity of the logarithm. □

Putting these estimates together with Equation (5.9) gives that for all $t > 0$,

$$\mathbb{P}\left[\left|\mathcal{N}_\theta^{(m)} - \frac{n\theta}{2\pi}\right| > t\right] \leq 2\exp\left(-\min\left\{\frac{t^2}{4m\left(\log\left(\frac{n}{m}\right) + 1\right)}, \frac{t}{2}\right\}\right). \tag{5.10}$$

This inequality gives a new route to eigenvalue rigidity: the individual eigenvalues tend to be very close to their predicted locations because of the concentration of the counting function. Rigidity of specific eigenvalues can be explicitly quantified as follows.

Lemma 5.21 *Let $1 \le m \le n$ and let $U \in \mathbb{U}(n)$ be uniformly distributed. Denote by $e^{i\theta_j}$, $1 \le j \le n$, the eigenvalues of U^m, ordered so that $0 \le \theta_1 \le \cdots \le \theta_n < 2\pi$. Then for each j and $u > 0$,*

$$\mathbb{P}\left[\left|\theta_j - \frac{2\pi j}{n}\right| > \frac{4\pi}{n}u\right] \le 4\exp\left[-\min\left\{\frac{u^2}{m\left(\log\left(\frac{n}{m}\right)+1\right)}, u\right\}\right]. \tag{5.11}$$

Proof For each $1 \le j \le n$ and $u > 0$, if $j + 2u < n$, then

$$\begin{aligned}\mathbb{P}\left[\theta_j > \frac{2\pi j}{n} + \frac{4\pi}{n}u\right] &= \mathbb{P}\left[\mathcal{N}^{(m)}_{\frac{2\pi(j+2u)}{n}} < j\right] = \mathbb{P}\left[j + 2u - \mathcal{N}^{(m)}_{\frac{2\pi(j+2u)}{n}} > 2u\right] \\ &\le \mathbb{P}\left[\left|\mathcal{N}^{(m)}_{\frac{2\pi(j+2u)}{n}} - \mathbb{E}\mathcal{N}^{(m)}_{\frac{2\pi(j+2u)}{n}}\right| > 2u\right].\end{aligned}$$

If $j + 2u \ge n$, then

$$\mathbb{P}\left[\theta_j > \frac{2\pi j}{n} + \frac{4\pi}{n}u\right] = \mathbb{P}\left[\theta_j > 2\pi\right] = 0,$$

and the above inequality holds trivially. The probability that $\theta_j < \frac{2\pi j}{n} - \frac{4\pi}{n}u$ is bounded in the same way. Inequality (5.11) now follows from (5.10). □

We are now in a position to bound the expected distance between the empirical spectral measure of U^m and uniform measure. Let θ_j be as in Lemma 5.21. Then by Fubini's theorem,

$$\begin{aligned}\mathbb{E}\left|\theta_j - \frac{2\pi j}{n}\right|^p &= \int_0^\infty pt^{p-1}\mathbb{P}\left[\left|\theta_j - \frac{2\pi j}{n}\right| > t\right] dt \\ &= \frac{(4\pi)^p p}{n^p}\int_0^\infty u^{p-1}\mathbb{P}\left[\left|\theta_j - \frac{2\pi j}{n}\right| > \frac{4\pi}{n}u\right] du \\ &\le \frac{4(4\pi)^p p}{n^p}\left[\int_0^\infty u^{p-1}e^{-u^2/m[\log(n/m)+1]}\, du + \int_0^\infty u^{p-1}e^{-u}\, du\right] \\ &= \frac{4(4\pi)^p}{n^p}\left[\left(m\left[\log\left(\frac{n}{m}\right)+1\right]\right)^{p/2}\Gamma\left(\frac{p}{2}+1\right) + \Gamma(p+1)\right] \\ &\le 8\Gamma(p+1)\left(\frac{4\pi}{n}\sqrt{m\left[\log\left(\frac{n}{m}\right)+1\right]}\right)^p.\end{aligned}$$

Let ν_n be the measure which puts mass $\frac{1}{n}$ at each of the points $e^{2\pi ij/n}$, $1 \le j \le n$. Then

$$
\begin{aligned}
\mathbb{E}W_p(\mu_{n,m}, \nu_n)^p &\le \mathbb{E}\left[\frac{1}{n}\sum_{j=1}^n \left|e^{i\theta_j} - e^{2\pi i j/n}\right|^p\right] \\
&\le \mathbb{E}\left[\frac{1}{n}\sum_{j=1}^n \left|\theta_j - \frac{2\pi j}{n}\right|^p\right] \\
&\le 8\Gamma(p+1)\left(\frac{4\pi}{n}\sqrt{m\left[\log\left(\frac{n}{m}\right)+1\right]}\right)^p \\
&\le Cp^{p+\frac{1}{2}}e^{-p}\left(\frac{4\pi}{n}\sqrt{m\left[\log\left(\frac{n}{m}\right)+1\right]}\right)^p,
\end{aligned}
$$

by Stirling's formula. It is easy to check that $W_p(\nu_n, \nu) \le \frac{\pi}{n}$, and thus

$$
\begin{aligned}
\mathbb{E}W_p(\mu_{n,m}, \nu) &\le \mathbb{E}W_p(\mu_{n,m}, \nu_n) + \frac{\pi}{n} \\
&\le \left(\mathbb{E}W_p(\mu_{n,m}, \nu_n)^p\right)^{\frac{1}{p}} + \frac{\pi}{n} \le \frac{Cp\sqrt{m\left[\log\left(\frac{n}{m}\right)+1\right]}}{n}.
\end{aligned}
\tag{5.12}
$$

The rest of the main theorem, namely the concentration of $W_p(\mu_{n,m}, \nu)$ at its mean, is a consequence of the concentration of measure phenomenon on the unitary group; the crucial point is that $W_p(\mu_{n,m}, \nu)$ is a Lipschitz function of U. The following lemma gives the necessary estimates.

Lemma 5.22 *Let $p \ge 1$. The map $A \mapsto \mu_A$ taking an $n \times n$ normal matrix to its spectral measure is Lipschitz with constant $n^{-1/\max\{p,2\}}$ with respect to W_p. In particular, if ρ is any fixed probability measure on $\mathbb{C}$, the map $A \mapsto W_p(\mu_A, \rho)$ is Lipschitz with constant $n^{-1/\max\{p,2\}}$.*

Proof If A and B are $n \times n$ normal matrices, then the Hoffman–Wielandt inequality [11, theorem VI.4.1] states that

$$
\min_{\sigma \in \Sigma_n} \sum_{j=1}^n \left|\lambda_j(A) - \lambda_{\sigma(j)}(B)\right|^2 \le \|A - B\|_{HS}^2, \tag{5.13}
$$

where $\lambda_1(A), \ldots, \lambda_n(A)$ and $\lambda_1(B), \ldots, \lambda_n(B)$ are the eigenvalues (with multiplicity, in any order) of A and B, respectively, and Σ_n is the group of permutations on n letters. Defining couplings of μ_A and μ_B given by

$$
\pi_\sigma = \frac{1}{n}\sum_{j=1}^n \delta_{(\lambda_j(A), \lambda_{\sigma(j)}(B))}
$$

for $\sigma \in \Sigma_n$, it follows from (5.13) that

$$\begin{aligned} W_p(\mu_A, \mu_B) &\leq \min_{\sigma \in \Sigma_n} \left(\frac{1}{n} \sum_{j=1}^{n} \left| \lambda_j(A) - \lambda_{\sigma(j)}(B) \right|^p \right)^{1/p} \\ &\leq n^{-1/\max\{p,2\}} \min_{\sigma \in \Sigma_n} \left(\sum_{j=1}^{n} \left| \lambda_j(A) - \lambda_{\sigma(j)}(B) \right|^2 \right)^{1/2} \\ &\leq n^{-1/\max\{p,2\}} \|A - B\|_{HS} . \end{aligned}$$

□

Now, by Theorem 3.14, $\mu_{n,m}$ is equal in distribution to the spectral measure of a block-diagonal $n \times n$ random matrix $U_1 \oplus \cdots \oplus U_m$, where the U_j are independent and uniform in $\mathbb{U}\left(\left\lfloor \frac{n}{m} \right\rfloor\right)$ and $\mathbb{U}\left(\left\lceil \frac{n}{m} \right\rceil\right)$. Identify $\mu_{n,m}$ with this measure and define the function $F(U_1, \ldots, U_m) = W_p(\mu_{U_1 \oplus \cdots \oplus U_m}, \nu)$; the preceding discussion means that if $U_1, \ldots, U_m$ are independent and uniform in $\mathbb{U}\left(\left\lfloor \frac{n}{m} \right\rfloor\right)$ and $\mathbb{U}\left(\left\lceil \frac{n}{m} \right\rceil\right)$ as necessary, then $F(U_1, \ldots, U_m) \overset{d}{=} W_p(\mu_{n,m}, \nu)$.

Applying the concentration inequality in Corollary 5.17 to the function F gives that

$$\mathbb{P}\left[F(U_1, \ldots, U_m) \geq \mathbb{E}F(U_1, \ldots, U_m) + t\right] \leq e^{-nt^2/24mL^2},$$

where L is the Lipschitz constant of F, and we have used the trivial estimate $\left\lfloor \frac{n}{m} \right\rfloor \geq \frac{n}{2m}$. Inserting the estimate of $\mathbb{E}F(U_1, \ldots, U_m)$ from Equation (5.12) and the Lipschitz estimates of Lemma 5.22 completes the proof of Theorem 5.18.

Notes and References

The general approach to concentration of measure taken here follows the writings of Michel Ledoux, in particular his book [71] and lecture notes [70]. I learned the proof of Theorem 5.9 from Nathaël Gozlan. The book [2] is a very accessible source for learning about log-Sobolev inequalities (if you read French). The book [13] by Boucheron, Lugosi, and Massart gives a more recent perspective with many applications.

A good reference for the basic notions of Riemannian geometry is [103]. Most of the exposition in Section 5.3 of the specific computations on the groups follows the corresponding exposition in the book [1] of Anderson, Guionnet, and Zeitouni. The fact that the curvature on (most of) the classical compact groups leads to sub-Gaussian concentration was first observed by Gromov and Milman [50], following earlier work of Gromov [49]; see also the appendix by Gromov in [83]. The Bakry–Émery Theorem first appeared in [8]; see also [9].

The coupling argument that circumvents the lack of a curvature bound on $\mathbb{U}(n)$ first appeared in [78].

The results of Section 5.4 are from [78], following earlier work in [77]. The use of the determinantal structure in obtaining bounds on expected Wasserstein distances was introduced by Dallaporta in [25, 26]. A survey of concentration of empirical spectral measures in many ensembles, including Haar measure on the classical compact groups, can be found in [74].

6

Geometric Applications of Measure Concentration

6.1 The Johnson–Lindenstrauss Lemma

An important area of application in computing is that of *dimension-reduction*. The essential problem is that in many settings, data sets live in very high-dimensional spaces. For example, a digital image can be encoded as a matrix, with each entry corresponding to one pixel, and the entry specifying the color of that pixel. That is, a small black-and-white image whose resolution was, say, 100×150 pixels would be encoded as a vector in $\{0, 1\}^{15,000}$. This presents a real problem because many algorithms for analyzing such high-dimensional data have their run-time increase very quickly as the dimension of the data increases, to the point that analyzing the data in the most obvious way becomes computationally infeasible – computer scientists refer to this as "the curse of dimensionality." The idea of dimension reduction is that in many situations, the desired algorithm can be at least approximately carried out in a much lower-dimensional setting than the one the data naturally lie in, which can make computationally infeasible problems feasible.

A Motivating Problem

Suppose you have a data set consisting of black-and-white images of hand-written examples of the numbers 1 and 2. That is, you have a library $\mathcal{X}$ of n points in $\mathbb{R}^d$, where d is the number of pixels in each image, with each point labeled (by a human) to indicate whether it is a 1 or a 2. You want to design a computer program so that one can input an image of a handwritten number, and the computer identifies it as a 1 or a 2. So the computer will have a query point $q \in \mathbb{R}^d$, and the natural thing to do is to program it to find the closest point in the library $\mathcal{X}$ to q; the computer then reports that the input image was of the same number as that closest point in $\mathcal{X}$.

The naïve approach would be for the computer to calculate the distance from q to each of the points of $\mathcal{X}$ in turn, keeping track of which point in $\mathcal{X}$ has so far been the closest. Such an algorithm runs in $O(nd)$ steps, which may be prohibitively many.

The idea of dimension reduction is to find a way to carry out the nearest point algorithm within some much lower-dimensional space, in such a way that you are guarranteed (or to be more realistic, very likely) to still find the closest point, *without* having to do much work to figure out which lower-dimensional space to work in. This sounds impossible, but the geometry of high-dimensional spaces often turns out to be surprising. The following important result about high-dimensional geometry has inspired many randomized algorithms incorporating dimension-reduction.

Lemma 6.1 (The Johnson–Lindenstrauss Lemma) *There are absolute constants c, C such that the following holds.*

Let $\{x_j\}_{j=1}^n \subseteq \mathbb{R}^d$, and let P be a random $k \times d$ matrix, consisting of the first k rows of a Haar-distributed random matrix in $\mathbb{SO}(d)$. Fix $\epsilon > 0$ and let $k = \frac{a\log(n)}{\epsilon^2}$. With probability $1 - Cn^{2-\frac{ac}{4}}$

$$(1-\epsilon)\|x_i - x_j\|^2 \leq \left(\frac{d}{k}\right)\|Px_i - Px_j\|^2 \leq (1+\epsilon)\|x_i - x_j\|^2 \tag{6.1}$$

for all $i, j \in \{1, \dots, n\}$.

That is, if n points in $\mathbb{R}^d$ are projected onto a random subspace of dimension on the order of $\log(n)$, then after appropriate rescaling, the pairwise distances between the points hardly changes. The practical conclusion of this is that if the application in question is about the metric structure of the data (finding the closest point as above, finding the most separated pair of points, finding the minimum length spanning tree of a graph, etc.), there is no need to work in the high-dimensional space that the data naturally live in, and that moreover there is no need to work hard to pick a lower-dimensional subspace onto which to project: a random one should do.

Getting an Almost-Solution, with High Probability

The discussion above suggests finding an approximate solution to the problem of finding the closest point to q in $\mathcal{X}$ by choosing a random $k \times d$ matrix P to be the first k rows of a Haar-distributed $U \in \mathbb{SO}(d)$, then finding the closest point in $\{Px :\in \mathcal{X}\}$ to Pq. There are two obvious causes for concern. One is that we might have the bad luck to choose a bad matrix P that doesn't satisfy (6.1). But that is *very* unlikely, and so typically one just accepts that risk and assumes it won't actually happen in practice.

There is a second issue, though, which is that it is possible to choose P that does satisfy (6.1), but to have the closest point in $\{Px :\in \mathcal{X}\}$ to Pq be Py, whereas the closest point in $\mathcal{X}$ to q is z, with $y \neq z$. In that case, although the approach above will yield the wrong identity for the closest point (y instead of z), it follows by choice of y and (6.1) that

$$\|q - y\| \leq \sqrt{\frac{d}{k(1-\epsilon)}}\|Pq - Py\| \leq \sqrt{\frac{d}{k(1-\epsilon)}}\|Pq - Pz\| \leq \sqrt{\frac{1+\epsilon}{1-\epsilon}}\|q - z\|.$$

So even though z is the true closest point to q, y is almost as close. In our example of recognizing whether a handwritten number is a 1 or a 2, it seems likely that even if we don't find the exact closest point in the reference set, the algorithm will still manage to correctly identify the number.

The payoff for being willing to accept an answer that may be not quite right and accepting the (tiny) risk that we'll choose a bad matrix is significant. The naïve algorithm mentioned at the beginning now runs in $O(n \log(n))$ steps rather than $O(nd)$ steps.

Proof of the Johnson–Lindenstrauss Lemma

Given $\{x_i\}_{i=1}^n \subseteq \mathbb{R}^d$, $\epsilon > 0$, and U a Haar-distributed random matrix in $\mathbb{SO}(d)$, let P be the $k \times d$ matrix consisting of the first k rows of U. The goal is to show that for each pair (i,j),

$$(1-\epsilon)\|x_i - x_j\|^2 \leq \left(\frac{d}{k}\right)\left\|Px_i - Px_j\right\|^2 \leq (1+\epsilon)\|x_i - x_j\|^2$$

with high probability, or equivalently,

$$\sqrt{1-\epsilon} \leq \sqrt{\frac{d}{k}}\left\|Px_{i,j}\right\| \leq \sqrt{1+\epsilon}$$

for $x_{i,j} := \frac{x_i - x_j}{\|x_i - x_j\|}$.

For notational convenience, fix i and j for the moment and let $x = x_{i,j}$. By the translation-invariance of Haar measure,

$$Px \stackrel{d}{=} Pe_1 = (U_{11}, \dots, U_{k1}),$$

where e_1 is the first standard basis vector in $\mathbb{R}^d$. Since the first column of U is distributed as a uniform random vector in $\mathbb{S}^{d-1}$, we may furthermore write

$$Px \stackrel{d}{=} (X_1, \dots, X_k),$$

with $X = (X_1, \dots, X_d)$ uniform in $\mathbb{S}^{d-1}$. Consider therefore the function $F : \mathbb{S}^{d-1} \to \mathbb{R}$ defined by

$$F(x_1, \dots, x_d) = \sqrt{\frac{d}{k}} \left\| (x_1, \dots, x_k) \right\| = \sqrt{\frac{d}{k}\left(x_1^2 + \cdots + x_k^2\right)}.$$

Let $x, y \in \mathbb{S}^{d-1}$; then

$$\begin{aligned} \left| F(x) - F(y) \right| &= \sqrt{\frac{d}{k}} \Big| \|(x_1, \dots, x_k)\| - \|(y_1, \dots, y_k)\| \Big| \\ &\leq \sqrt{\frac{d}{k}} \left\| (x_1 - y_1, \dots, x_k - y_k) \right\| \leq \sqrt{\frac{d}{k}} \|x - y\|. \end{aligned}$$

That is, the function F is $\sqrt{\frac{d}{k}}$-Lipschitz on $\mathbb{S}^{d-1}$, and so concentration of measure on the sphere (i.e., Lévy's lemma) applies:

$$\mathbb{P}\left[|F(X) - \mathbb{E}F(X)| \geq \epsilon \right] \leq Ce^{-ck\epsilon^2}. \tag{6.2}$$

To complete the proof, it remains to show that $\mathbb{E}F(X) \approx 1$.

Since $\mathbb{E}X_i^2 = \frac{1}{d}$ for each i, $\mathbb{E}\left[F^2(X)\right] = 1$; written slightly differently,

$$1 = \mathrm{Var}(F(X)) + \left(\mathbb{E}F(X)\right)^2.$$

By Fubini's theorem and the concentration inequality (6.2),

$$\mathrm{Var}(F(X)) = \int_0^\infty \mathbb{P}\left[|F(X) - \mathbb{E}F(X)|^2 \geq t \right] dt \leq \int_0^\infty Ce^{-ckt} dt = \frac{C}{ck},$$

so that

$$\sqrt{1 - \frac{C}{ck}} \leq \mathbb{E}F(X) \leq 1.$$

Recall that $k = \frac{a \log(n)}{\epsilon^2}$. As long as $\epsilon < \frac{ca \log(n)}{C + ca \log(n)}$, this means that $1 - \frac{\epsilon}{2} \leq \mathbb{E}F(X) \leq 1$, and so

$$\mathbb{P}\left[\left| F(X) - 1 \right| > \epsilon \right] \leq Ce^{-\frac{ck\epsilon^2}{4}}; \tag{6.3}$$

that is, with probability at least $1 - Ce^{-\frac{ck\epsilon^2}{4}}$,

$$1 - \epsilon \leq \sqrt{\frac{d}{k}} \|Px\| \leq 1 + \epsilon.$$

Returning to the original formulation, for each pair (i, j), there is a set of probability at least $1 - Ce^{-\frac{ck\epsilon^2}{4}}$ such that

$$(1-\epsilon)^2 \|x_i - x_j\|^2 \leq \left(\frac{d}{k}\right) \left\|Px_i - Px_j\right\|^2 \leq (1+\epsilon)^2 \|x_i - x_j\|^2.$$

There are fewer than n^2 pairs (i, j), so a simple union bound gives that the above statement holds *for all pairs* (i, j) with probability at least $1 - \frac{C}{n^{\frac{ac}{4}-2}}$. □

6.2 Dvoretzky's Theorem

The following theorem is one of the foundational results of the local theory of Banach spaces; V. Milman's proof [82] gave the first explicit use of the concentration of measure phenomenon in Banach space theory.

Theorem 6.2 (Dvoretzky's theorem) *Let $\|\cdot\|$ be an arbitrary norm on $\mathbb{C}^n$. There is an invertible linear map $T : \mathbb{R}^n \to \mathbb{R}^n$ such that for all $\epsilon > 0$, if $k \leq C\epsilon^2 \log(n)$ and if $E \subseteq \mathbb{R}^n$ is a random k-dimensional subspace of $\mathbb{R}^n$, then with probability at least $1 - e^{-ck}$,*

$$1 - \epsilon \leq \frac{\|Tv\|}{|v|} \leq 1 + \epsilon \quad \textit{for all } v \in E,$$

where c, C are absolute constants, independent of $\|\cdot\|$, and $|\cdot|$ denotes the Euclidean norm.

The phrase "random k-dimensional subspace" in the statement of the theorem should be understood as in the previous section: as the linear span of the first k columns of U, where U is distributed according to Haar measure on $\mathbb{U}(n)$. The distribution of such a random subspace is the unique probability measure on the Grassmannian $\mathfrak{G}_{n,k}^{\mathbb{C}}$ of k-dimensional subspaces of $\mathbb{C}^n$, which is invariant under the action of $\mathbb{U}(n)$.

Milman's proof of Dvoretzky's theorem used the concentration of measure on the sphere, but using the more recently proved concentration of measure on the unitary group, one can deduce the theorem more directly. Before proceeding with the main body of the proof, we make a simple geometric reduction. Throughout this section, $\|\cdot\|$ will denote the arbitrary norm in the statement of the theorem, and $|\cdot|$ will denote the Euclidean norm.

Recall that an **ellipsoid** is defined to be a linear image of the Euclidean unit ball $\mathbb{S}^n_{\mathbb{C}}$. By applying an initial linear transformation, it suffices to assume that the ellipsoid of maximal volume contained in the unit ball of the norm $\|\cdot\|$ is $\mathbb{S}^n_{\mathbb{C}}$ itself. This implies in particular that

$$\|v\| \leq |v| \quad \text{for all } v \in \mathbb{C}^n.$$

Our approach to Theorem 6.2 centers around the random quantity

$$X_v(U) := \|Uv\| - \mathbb{E}\|Uv\|,$$

where $v \in \mathbb{C}^n$ is a fixed unit vector, and U is a random unitary matrix. In particular, for a subspace $E \subseteq \mathbb{C}^n$, the supremum

$$\sup_{v \in E \cap \mathbb{S}^n_{\mathbb{C}}} |X_v(U)|$$

measures the (random) variability of the quantity $\|Uv\|$ over $v \in E$.

Proposition 6.3 *For $v \in \mathbb{S}^n_{\mathbb{C}}$, let $X_v(U) = \|Uv\| - \mathbb{E}\|Uv\|$, with U a Haar-distributed random unitary matrix. Let $E \subseteq \mathbb{C}^n$ be any subspace. Then*

$$\mathbb{P}\left[\left|\sup_{v \in E \cap \mathbb{S}^n_{\mathbb{C}}} |X_v(U)| - \mathbb{E}\left(\sup_{v \in E \cap \mathbb{S}^n_{\mathbb{C}}} |X_v(U)|\right)\right| > t\right] \leq Ce^{-cnt^2}.$$

Proof Let $U, U' \in \mathbb{U}(n)$. Then

$$\begin{aligned}
\left|\sup_v |X_v(U)| - \sup_v |X_v(U')|\right| &\leq \sup_v \big||X_v(U)| - |X_v(U')|\big| \\
&= \sup_v \Big|\big|\|Uv\| - \mathbb{E}\|Uv\|\big| - \big|\|Uv\| - \mathbb{E}\|U'v\|\big|\Big| \\
&\leq \sup_v \|(U - U')v\| \\
&\leq \|U - U'\|_{H.S.},
\end{aligned}$$

making use of the facts that $\|\cdot\| \leq |\cdot|$ and $\|\cdot\|_{op} \leq \|\cdot\|_{H.S.}$.

The function $U \mapsto \sup_v |X_v(U)|$ is thus 1-Lipschitz, and the result follows from Theorem 5.17. □

The random quantity $\sup_{v \in E \cap \mathbb{S}^n_{\mathbb{C}}} |X_v(U)|$ is thus typically close to its mean; the following lemma is the main ingredient needed to estimate that mean.

Lemma 6.4 *Let $x, y \in \mathbb{S}^n_{\mathbb{C}}$ with $x \neq y$, and let U be a random $n \times n$ unitary matrix. Then for all $t > 0$,*

$$\mathbb{P}\left[\big|\|Ux\| - \|Uy\|\big| > t\right] \leq Ce^{-\frac{cnt^2}{|x-y|^2}}.$$

Proof First, note that $\|U(e^{i\theta}y)\| = \|Uy\|$ for any θ. Choosing θ such that $\langle x, e^{i\theta}y\rangle$ is real and nonnegative means that $\operatorname{Re}\left(\langle x, e^{i\theta}y\rangle\right) \geq \operatorname{Re}\left(\langle x, y\rangle\right)$, and so

$$|x - e^{i\theta}y|^2 = |x|^2 + |y|^2 - 2\operatorname{Re}\left(\left\langle x, e^{i\theta}y\right\rangle\right) \leq |x - y|^2.$$

We may therefore assume that $\langle x, y\rangle$ is real. Let $z := \frac{x+y}{2}$ and $w := \frac{x-y}{2}$, so that

$$x = z + w \qquad y = z - w \qquad z \perp w;$$

in terms of z and w, the desired conclusion is

$$\mathbb{P}\left[\big|\|Uz + Uw\| - \|Uz - Uw\|\big| > t\right] \leq Ce^{-\frac{cnt^2}{|w|^2}}.$$

By the translation-invariance of Haar measure, it suffices to assume that $z = e_1$ (the first standard basis vector). Observe that, conditional on the event $\{Ue_1 = u\}$, the distribution of Uw is the same as that of $-Uw$: conditioning on $Ue_1 = u$ simply means choosing the first column of U to be u, and filling out the rest of the matrix column by column as described in Section 1.2. In particular, the conditional distribution of U given $Ue_1 = u$ is invariant under changing the sign of each of the remaining columns; doing so replaces Uw by $-Uw$. It follows that

$$\mathbb{E}\left[\|u + Uw\| - \|u - Uw\|\middle|Ue_1 = u\right] = 0.$$

Moreover, if $U \in \mathbb{U}(n)$ with $Ue_1 = u$, the function $U \mapsto \|u + Uw\|$ is a $|w|$-Lipschitz function of the remaining columns: if U' is another such matrix, then

$$\Big|\|u + Uw\| - \|u + U'w\|\Big| \leq \|(U - U')w\| \leq |(U - U')w| \leq \|U - U'\|_{H.S.}|w|,$$

again using that $\|\cdot\| \leq |\cdot|$ and $\|\cdot\|_{op} \leq \|\cdot\|_{H.S.}$.

The conditional version of the column-by-column construction of Haar measure described above makes it clear that conditional on $Ue_1 = u$, the rest of the matrix U is distributed according to Haar measure on a copy of $\mathbb{U}(n-1)$ embedded in $\mathbb{U}(n)$, and so Theorem 5.17 applies to give that for each u,

$$\mathbb{P}\left[\big|\|Uz + Uw\| - \|Uz - Uw\|\big| > t\middle|Ue_1 = u\right] \leq Ce^{-\frac{cnt^2}{|w|^2}}.$$

Averaging over u completes the proof. □

Proposition 6.5 *Let $E \subseteq \mathbb{C}^n$ be a subspace of dimension k. Then*

$$\mathbb{E}\sup_{v \in E \cap \mathbb{S}^n_{\mathbb{C}}} |X_v| \leq C\sqrt{\frac{k}{n}}.$$

Proof Lemma 6.4 shows exactly that the stochastic process $\{X_v\}_{v\in E\cap\mathbb{S}^n_{\mathbb{C}}}$ is sub-Gaussian, with respect to the metric $\frac{|\cdot|}{\sqrt{n}}$. Under a sub-Gaussian increment condition, Dudley's entropy bound (see, e.g., [100]) gives that

$$\begin{aligned}\mathbb{E}\sup_{v\in E\cap\mathbb{S}^n_{\mathbb{C}}}|X_v| &\le C\int_0^\infty\sqrt{\log\left(\mathcal{N}\left(E\cap\mathbb{S}^n_{\mathbb{C}},\frac{|\cdot|}{\sqrt{n}},\epsilon\right)\right)}d\epsilon\\ &=\frac{C}{\sqrt{n}}\int_0^\infty\sqrt{\log\left(\mathcal{N}\left(E\cap\mathbb{S}^n_{\mathbb{C}},|\cdot|,\epsilon\right)\right)}d\epsilon,\end{aligned}$$

where the covering number $\mathcal{N}\left(E\cap\mathbb{S}^n_{\mathbb{C}},|\cdot|,\epsilon\right)$ is the number of ϵ-balls needed to cover $E\cap\mathbb{S}^n_{\mathbb{C}}$, with respect to the distance $|\cdot|$. In particular, the integrand is zero for $\epsilon>2$. The covering number can be bounded using a simple volume argument (see lemma 2.6 of [83]) by $\exp\left(k\log\left(\frac{3}{\epsilon}\right)\right)$, and this completes the proof. □

Combining Propositions 6.3 and 6.5 gives that if E is a k-dimensional subspace of $\mathbb{C}^n$, then with probability at least $1-Ce^{-cnt^2}$,

$$\Big|\|Uv\|-\mathbb{E}\|Uv\|\Big|\le t+C\sqrt{\frac{k}{n}}$$

for all $v\in E$; the next step of the proof of Theorem 6.2 is to estimate $\mathbb{E}\|Uv\|$.

Proposition 6.6 *There is a universal constant c such that for $v\in\mathbb{C}^n$ with $|v|=1$, U a random unitary matrix, and $\|\cdot\|$ as above,*

$$c\sqrt{\frac{\log(n)}{n}}\le\mathbb{E}\|Uv\|\le 1.$$

Proof The upper bound is trivial, since $\|\cdot\|\le|\cdot|$.

For the lower bound, it follows from the Dvoretzky-Rogers lemma and its proof (see Lemma 3.16 of [71]) that, under the condition on $\|\cdot\|$ discussed at the beginning of the section (i.e., that the maximum-volume ellipsoid contained in the unit ball of $\|\cdot\|$ is in fact the Euclidean unit ball), there is an orthonormal basis $\{v_1,\ldots,v_n\}$ of $\mathbb{R}^n$ such that

$$\|v_j\|\ge\frac{1}{2},\quad 1\le j\le\left\lfloor\frac{n}{2}\right\rfloor;$$

by applying a further linear isometry, we assume that this estimate holds for the standard basis $\{e_i\}$.

Now, $Uv\overset{d}{=}X$, where X is uniformly distributed on $\mathbb{S}^n_{\mathbb{C}}$. Moreover, if $(\epsilon_1,\ldots,\epsilon_n)$ is a random vector of i.i.d. centered $\{-1,1\}$-valued random variables, then $X \overset{d}{=} (\epsilon_1 X_1, \ldots, \epsilon_n X_n)$. For fixed j, conditional on X and ϵ_j, it follows by Jensen's inequality that

$$|X_j| \|e_j\| = \left\| \mathbb{E}\left[(\epsilon_1 X_1, \ldots, \epsilon_n X_n) \middle| \epsilon_j, X\right] \right\| \leq \mathbb{E}\left[\|(\epsilon_1 X_1, \ldots, \epsilon_n X_n)\| \;\middle|\; \epsilon_j, X\right].$$

Averaging over ϵ_j gives

$$|X_j| \|e_j\| \leq \mathbb{E}\left[\|(\epsilon_1 X_1, \ldots, \epsilon_n X_n)\| \;\middle|\; X\right].$$

Taking the maximum over $j \in \{1, \ldots, m\}$ and then taking expectation of both sides gives

$$\mathbb{E}\left[\max_{1 \leq j \leq m} |X_j| \|e_j\|\right] \leq \mathbb{E}\|(\epsilon_1 X_1, \ldots, \epsilon_n X_n)\| = \mathbb{E}\|X\|,$$

and so

$$\mathbb{E}\|X\| \geq \frac{1}{2} \mathbb{E}\left[\max_{1 \leq j \leq m} |X_j|\right], \tag{6.4}$$

where $m := \left\lfloor \frac{n}{2} \right\rfloor$. To estimate the right-hand side of (6.4), we use a standard trick of relating the spherical expectation to a corresponding Gaussian one. Note that a uniform random vector on $\mathbb{S}^n_{\mathbb{C}}$ can be naturally identified with a uniform random vector on $\mathbb{S}^{2n}$, so there is no loss in considering the real case. (That the uniform measure on $\mathbb{S}^n_{\mathbb{C}}$ is mapped to uniform measure on $\mathbb{S}^{2n}$ by the obvious identification is slightly nontrivial but follows by the uniqueness of Haar measure and the fact that $\mathbb{U}(n)$ acts transitively on $\mathbb{S}^n_{\mathbb{C}}$.)

Let $\{Z_1, \ldots, Z_n\}$ be i.i.d. standard Gaussian random variables. Then it is well known that there is a constant ρ such that

$$\mathbb{E}\left[\max_{1 \leq j \leq m} |Z_j|\right] \geq 2\rho\sqrt{\log(m)}.$$

Writing the expectation above explicitly in terms of the density of the Z_j in spherical coordinates gives

$$\begin{aligned}
\mathbb{E}\left[\max_{1 \leq j \leq m} |Z_j|\right] &= \frac{1}{(2\pi)^{m/2}} \int_{\mathbb{S}^{m-1}} \int_0^\infty \max_{1 \leq j \leq m} |r y_j| e^{-r^2/2} r^{m-1} dr d\sigma(y) \\
&= \frac{\Gamma\left(\frac{m+1}{2}\right)}{\sqrt{2}\pi^{m/2}} \int_{\mathbb{S}^{m-1}} \max_{1 \leq j \leq m} |y_j| d\sigma(y),
\end{aligned}$$

where σ denotes the surface area measure on $\mathbb{S}^{m-1}$. The surface area of the unit sphere in $\mathbb{R}^m$ is $\frac{m\pi^{m/2}}{\Gamma(\frac{m}{2}+1)}$, and so rearranging above and applying the lower bound (6.4) gives

$$\mathbb{E}\left[\max_{1\leq j\leq m}|X_j|\right] \geq \frac{2\sqrt{2}\rho\sqrt{\log(m)}\Gamma\left(\frac{m}{2}+1\right)}{m\Gamma\left(\frac{m+1}{2}\right)} \geq c\sqrt{\frac{\log(m)}{m}},$$

where the last estimate follows from Stirling's formula. □

We are now in a position to complete the proof of Dvoretzky's theorem. Let $M := \mathbb{E}\|Ue_1\|$, and for $\epsilon > 0$ fixed, let $k = cnM^2\epsilon^2$, where c is a small (but universal) constant. Applying Proposition 6.3 with $t = \frac{M\epsilon}{2}$ gives that for any k-dimensional subspace $E \subseteq \mathbb{C}^n$, with probability at least $1 - Ce^{-cn\epsilon^2M^2} \geq 1 - Ce^{-c\epsilon^2\log(n)}$,

$$M(1-\epsilon) \leq \|Uv\| \leq M(1+\epsilon),$$

for all $v \in E$ with $|v| = 1$. In particular, let E be the span of $\{e_1, \ldots, e_k\}$. Then the statement above means that with probability at least $1 - Ce^{-c\epsilon^2\log(n)}$,

$$1-\epsilon \leq \frac{\frac{1}{M}\|w\|}{|w|} \leq 1+\epsilon$$

for all w in the linear span of the first k columns of U; that is, for all w in a randomly chosen k-dimensional subspace of $\mathbb{C}^n$. Absorbing the constant M into the linear map T in the statement of Theorem 6.2 completes the proof.

6.3 A Measure-Theoretic Dvoretzky Theorem

In this section, the objects of study are the marginal distributions of high-dimensional probability measures. It was recognized long ago that in many settings, most one-dimensional projections of high-dimensional probability measures are approximately Gaussian. In particular, Borel's lemma (Lemma 2.4) is an early example: all one-dimensional projections of uniform measure on the sphere in $\mathbb{R}^d$ are the same, and all are approximately Gaussian for large d. This is also a familiar phenomenon in statistics, in which low-dimensional projections of high-dimensional data that appear approximately Gaussian are usually regarded as not giving useful information about the structure of the data.

It is natural to ask under what conditions on the high-dimensional distribution this phenomenon occurs, and moreover, how long it persists; i.e., if a d-dimensional probability distribution is projected onto a k-dimensional subspace, how large can k be relative to d so that such projections are typically approximately Gaussian?

The connection to Dvoretzky's theorem is the following. In both settings, an additional structure is imposed on $\mathbb{R}^n$ (a norm in the case of Dvoretzky's theorem; a probability measure in the present context); in either case, there is a particularly nice way to do this (the Euclidean norm and the Gaussian distribution, respectively). The question is then: If one projects an arbitrary norm or probability measure onto lower dimensional subspaces, does it tend to resemble this nice structure? If so, by how much must one reduce the dimension in order to see this phenomenon?

Theorem 6.7 *Let X be a random vector in $\mathbb{R}^d$ satisfying*

$$\mathbb{E}|X|^2 = \sigma^2 d \qquad \mathbb{E}\left||X|^2\sigma^{-2} - d\right| \leq L\frac{d}{(\log d)^{1/3}} \qquad \sup_{\xi \in \mathbb{S}^{d-1}} \mathbb{E}\,\langle \xi, X\rangle^2 \leq 1.$$

Let $X_{\mathbf{V}}^{(k)} := \big(\langle X, \mathbf{V}_1\rangle, \ldots, \langle X, \mathbf{V}_k\rangle\big)$, where

$$\mathbf{V} = \begin{bmatrix} - & \mathbf{V}_1 & - \\ & \vdots & \\ - & \mathbf{V}_d & - \end{bmatrix}$$

is an orthogonal matrix. Fix $\delta < 2$ and suppose that $k = \delta\frac{\log(d)}{\log(\log(d))}$. Then there is a $c > 0$ depending only on δ, such that for $\epsilon = \exp\left[-c\log(\log(d))\right]$, there is a subset $\mathfrak{T} \subseteq \mathbb{SO}(d)$ with $\mathbb{P}[\mathfrak{T}] \geq 1 - C\exp\left(-c'd\epsilon^2\right)$, such that for all $\mathbf{V} \in \mathfrak{T}$,

$$d_{BL}(X_{\mathbf{V}}^{(k)}, \sigma Z) \leq C'\epsilon.$$

The following example shows that, without additional assumptions, the theorem gives the best possible estimate on k.

Let X be distributed uniformly among $\{\pm\sqrt{d}e_1, \ldots, \pm\sqrt{d}e_d\}$, where the e_i are the standard basis vectors of $\mathbb{R}^d$. That is, X is uniformly distributed on the vertices of a cross-polytope. Then $\mathbb{E}[X] = 0$, $|X|^2 \equiv d$, and given $\xi \in \mathbb{S}^{d-1}$, $\mathbb{E}\,\langle X, \xi\rangle^2 = 1$, thus Theorem 6.7 applies.

Consider a projection of $\{\pm\sqrt{d}e_1, \ldots, \pm\sqrt{d}e_d\}$ onto a random subspace E of dimension k, and define the Lipschitz function $f : E \to \mathbb{R}$ by $f(x) := (1 - d(x, S_E))_+$, where S_E is the image of $\{\pm\sqrt{d}e_1, \ldots, \pm\sqrt{d}e_d\}$ under projection onto E and $d(x, S_E)$ denotes the (Euclidean) distance from the point x to the set S_E. Then if μ_{S_E} denotes the probability measure putting equal mass at each of the points of S_E, $\int f d\mu_{S_E} = 1$. On the other hand, the volume ω_k of the unit ball in $\mathbb{R}^k$ is asymptotically given by $\frac{\sqrt{2}}{\sqrt{k\pi}}\left[\frac{2\pi e}{k}\right]^{\frac{k}{2}}$ for large k, in the sense that the ratio tends to one as k tends to infinity. It follows that the standard Gaussian measure of a ball of radius 1 in $\mathbb{R}^k$ is bounded by $\frac{1}{(2\pi)^{k/2}}\omega_k \sim \frac{\sqrt{2}}{\sqrt{k\pi}}\left[\frac{e}{k}\right]^{\frac{k}{2}}$. If γ_k denotes the standard Gaussian measure in $\mathbb{R}^k$, then this estimate means that

$\int f d\gamma_k \leq \frac{2\sqrt{2}d}{\sqrt{k\pi}} \left[\frac{e}{k}\right]^{\frac{k}{2}}$. Now, if $k = \frac{c\log(d)}{\log(\log(d))}$ for $c > 2$, then this bound tends to zero, and thus $d_{BL}(\mu_{S_E}, \gamma_k)$ is close to 1 for any choice of the subspace E; the measures μ_{S_E} are far from Gaussian in this regime.

This example together with Theorem 6.7 show that the phenomenon of typically Gaussian marginals persists for $k = \frac{c\log(d)}{\log(\log(d))}$ for $c < 2$ but fails in general if $k = \frac{c\log(d)}{\log(\log(d))}$ for $c > 2$.

The proof of Theorem 6.7 is in several steps. Borrowing terminology from statistical mechanics, we first consider the "annealed" version of $X_{\mathbf{V}}^{(k)}$, in which $\mathbf{V}$ is taken to be random and independent of X, and show that it is approximately Gaussian. Then, we show that the random distance between a "quenched" version of $X_{\mathbf{V}}^{(k)}$ and its annealed (i.e., averaged) version is strongly concentrated at its mean. Finally, we estimate this "average distance to average".

Theorem 6.8 *Let X be a random vector in $\mathbb{R}^n$, with $\mathbb{E}X = 0$, $\mathbb{E}\left[|X|^2\right] = \sigma^2 d$, and*

$$\mathbb{E}\left||X|^2\sigma^{-2} - d\right| := A < \infty.$$

Suppose that $\mathbf{V}$ is distributed according to Haar measure on $\mathbb{SO}(d)$ and independent of X, and let $X_{\mathbf{V}}^{(k)} = \big(\langle X, \mathbf{V}_1\rangle, \dots, \langle X, \mathbf{V}_k\rangle\big)$, where $\mathbf{V}_i$ is the ith row of $\mathbf{V}$. Then

$$d_{BL}(X_{\mathbf{V}}^{(k)}, \sigma Z) \leq \frac{\sigma\sqrt{k}(A+1) + \sigma k}{d-1}.$$

Proof The proof is via the version of Stein's method given in Theorem 2.21, Section 2.4. Observe first that $\mathbb{E}X_{\mathbf{V}}^{(k)} = 0$ by symmetry and if v_{ij} denotes the i-j^{th} entry of $\mathbf{V}$, then

$$\begin{aligned}\mathbb{E}(X_{\mathbf{V}}^{(k)})_i(X_{\mathbf{V}}^{(k)})_j &= \mathbb{E}\langle \mathbf{V}_i, X\rangle \big\langle \mathbf{V}_j, X\big\rangle = \sum_{r,s=1}^{d} \mathbb{E}\left[v_{ir}v_{js}\right]\mathbb{E}\left[X_r X_s\right] \\ &= \frac{\delta_{ij}}{d}\mathbb{E}\left[|X|^2\right] = \delta_{ij}\sigma^2,\end{aligned}$$

thus $\mathbb{E}[X_{\mathbf{V}}^{(k)}(X_{\mathbf{V}}^{(k)})^T] = \sigma^2 I_d$.

The construction of $X_{\mathbf{V},\epsilon}$ for the application Theorem 2.21 is analogous to the construction of exchangeable pairs of random matrices in Section 2.4. Let

$$\mathbf{A}_\epsilon := \begin{bmatrix} \sqrt{1-\epsilon^2} & \epsilon \\ -\epsilon & \sqrt{1-\epsilon^2} \end{bmatrix} \oplus \mathbf{I}_{d-2} = \mathbf{I}_d + \begin{bmatrix} -\frac{\epsilon^2}{2} + \delta & \epsilon \\ -\epsilon & -\frac{\epsilon^2}{2} + \delta \end{bmatrix} \oplus \mathbf{0}_{d-2},$$

where $\delta = O(\epsilon^4)$. Let $\mathbf{U} \in \mathbb{SO}(d)$ be a random orthogonal matrix, independent of X and $\mathbf{V}$, and define

$$X_{\mathbf{V},\epsilon} := \left(\left\langle \mathbf{U}\mathbf{A}_\epsilon \mathbf{U}^T \mathbf{V}_1, X \right\rangle, \dots, \left\langle \mathbf{U}\mathbf{A}_\epsilon \mathbf{U}^T \mathbf{V}_k, X \right\rangle \right);$$

the pair $(X_{\mathbf{V}}^{(k)}, X_{\mathbf{V},\epsilon})$ is exchangeable by the rotation invariance of the distribution of $\mathbf{V}$, and so $X_{\mathbf{V}}^{(k)} \stackrel{d}{=} X_{\mathbf{V},\epsilon}$.

Let $\mathbf{K}$ be the $d \times 2$ matrix given by the first two columns of $\mathbf{U}$ and let $\mathbf{C} = \begin{bmatrix} 0 & 1 \\ -1 & 0 \end{bmatrix}$; define the matrix $\mathbf{Q} = \left[q_{ij} \right]_{i,j=1}^d = \mathbf{K}\mathbf{C}\mathbf{K}^T$. Then

$$\mathbf{U}\mathbf{A}_\epsilon \mathbf{U}^T - \mathbf{I}_d = \left(-\frac{\epsilon^2}{2} + \delta \right) \mathbf{K}\mathbf{K}^T + \epsilon \mathbf{Q},$$

and so, writing $X_{\mathbf{V}}^{(k)} = (X_1^{\mathbf{V}}, \dots, X_k^{\mathbf{V}})$ and $X_{\mathbf{V},\epsilon} = (X_{\epsilon,1}^{\mathbf{V}}, \dots, X_{\epsilon,k}^{\mathbf{V}})$,

$$\begin{aligned} \mathbb{E}\left[X_{\epsilon,j}^{\mathbf{V}} - X_j^{\mathbf{V}} \middle| X, \mathbf{V} \right] &= \mathbb{E}\left[\left\langle (\mathbf{U}\mathbf{A}_\epsilon \mathbf{U}^T - \mathbf{I}_d)\mathbf{V}_j, X \right\rangle \middle| X, \mathbf{V} \right] \\ &= \epsilon \mathbb{E}\left[\langle \mathbf{Q}\mathbf{V}_j, X \rangle \middle| X, \mathbf{V} \right] - \left(\frac{\epsilon^2}{2} + \delta \right) \mathbb{E}\left[\left\langle \mathbf{K}\mathbf{K}^T \mathbf{V}_j, X \right\rangle \middle| X, \mathbf{V} \right]. \end{aligned}$$

Recall that $\mathbf{Q}$ and $\mathbf{K}$ are determined by $\mathbf{U}$ alone, and that $\mathbf{U}$ is independent of $X, \mathbf{V}$. It is easy to show that $\mathbb{E}\left[\mathbf{Q}\right] = \mathbf{0}_d$ and $\mathbb{E}\left[\mathbf{K}\mathbf{K}^T\right] = \frac{2}{d}\mathbf{I}_d$, thus

$$\mathbb{E}\left[X_{\mathbf{V},\epsilon} - X_{\mathbf{V}}^{(k)} \middle| X, \mathbf{V} \right] = \left(-\frac{\epsilon^2}{d} + \frac{2\delta}{d} \right) X_{\mathbf{V}}^{(k)}.$$

Condition 1 of Theorem 2.21 is thus satisfied with $\lambda(\epsilon) = \frac{\epsilon^2}{d}$.

It follows from the formula given in Lemma 2.22 that $\mathbb{E} q_{rs} q_{tw} = \frac{2}{d(d-1)} \left[\delta_{rt}\delta_{sw} - \delta_{rw}\delta_{st} \right]$, which yields

$$\begin{aligned} &\mathbb{E}\left[(X_{\epsilon,j}^{\mathbf{V}} - X_j^{\mathbf{V}})(X_{\epsilon,\ell}^{\mathbf{V}} - X_\ell^{\mathbf{V}}) \middle| X, \mathbf{V} \right] \\ &\quad = \epsilon^2 \mathbb{E}\left[\langle \mathbf{Q}\mathbf{V}_j, X \rangle \langle \mathbf{Q}\mathbf{V}_\ell, X \rangle \middle| X, \mathbf{V} \right] + O(\epsilon^3) \\ &\quad = \epsilon^2 \sum_{r,s,t,w=1}^d \mathbb{E}\left[q_{rs} q_{tw} v_{js} v_{\ell w} X_r X_t \middle| X, \mathbf{V} \right] + O(\epsilon^3) \\ &\quad = \frac{2\epsilon^2}{d(d-1)} \left[\sum_{r,s=1}^d v_{js} v_{\ell s} X_r^2 - \sum_{r,s=1}^d v_{js} v_{\ell r} X_r X_s \right] + O(\epsilon^3) \end{aligned}$$

$$= \frac{2\epsilon^2}{d(d-1)}\left[\delta_{j\ell}|X|^2 - X_j^{\mathbf{V}}X_\ell^{\mathbf{V}}\right] + O(\epsilon^3)$$

$$= \frac{2\epsilon^2\sigma^2}{d}\delta_{j\ell} + \frac{2\epsilon^2}{d(d-1)}\left[\delta_{j\ell}\left(|X|^2 - \sigma^2 d\right) + \delta_{j\ell}\sigma^2 - X_j^{\mathbf{V}}X_\ell^{\mathbf{V}}\right] + O(\epsilon^3).$$

The random matrix $\mathbf{F}$ of Theorem 2.21 is therefore defined by

$$\mathbf{F} = \frac{1}{d-1}\left[\left(|X|^2 - \sigma^2 d\right)\mathbf{I}_k + \sigma^2\mathbf{I}_k - X_{\mathbf{V}}^{(k)}(X_{\mathbf{V}}^{(k)})^T\right].$$

It now follows from Theorem 2.21 that

$$\begin{aligned} d_{BL}(X_{\mathbf{V}}^{(k)}, \sigma Z) &\leq W_1(X_{\mathbf{V}}^{(k)}, \sigma Z) \\ &\leq \frac{1}{\sigma}\mathbb{E}\|F\|_{H.S.} \\ &\leq \frac{\sigma\sqrt{k}}{d-1}\left[\mathbb{E}\left|\frac{|X|^2}{\sigma^2} - d\right| + 1\right] + \frac{\sigma}{d-1}\mathbb{E}\left[\sum_{j=1}^k \left(\frac{X_j^{\mathbf{V}}}{\sigma}\right)^2\right] \qquad (6.5) \\ &\leq \frac{\sigma\sqrt{k}(A+1) + \sigma k}{d-1}. \end{aligned}$$

□

The next result gives the concentration of $d_{BL}(X_{\mathbf{V}}^{(k)}, \sigma Z)$ about its mean. The idea is very similar to the argument at the end of Section 5.4 on the concentration of the empirical spectral measure.

Theorem 6.9 *Let $X \in \mathbb{R}^d$ be a centered random vector, with $\mathbb{E}\left[|X|^2\right] = \sigma^2 d$, and let*

$$B := \sup_{\xi \in \mathbb{S}^{d-1}} \mathbb{E}\langle X, \xi\rangle^2.$$

The function

$$\mathbf{V} \longmapsto d_{BL}(X_{\mathbf{V}}^{(k)}, \sigma Z)$$

on $\mathbb{SO}(d)$ can be viewed as a random variable, by letting $\mathbf{V}$ be distributed according to Haar measure on $\mathbb{SO}(d)$. Then there are universal constants C, c such that for any $\epsilon > 0$,

$$\mathbb{P}\left[\left|d_{BL}(X_{\mathbf{V}}^{(k)}, \sigma Z) - \mathbb{E}d_{BL}(X_{\mathbf{V}}^{(k)}, \sigma Z)\right| > \epsilon\right] \leq Ce^{-\frac{cd\epsilon^2}{B}}.$$

Proof Define a function $F : \mathbb{SO}(d) \to \mathbb{R}$ by

$$F(\mathbf{V}) = \sup_{\|f\|_{BL}\leq 1}\left|\mathbb{E}_X f(X_{\mathbf{V}}^{(k)}) - \mathbb{E}f(\sigma Z)\right|,$$

where $\mathbb{E}_X$ denotes the expectation with respect to the distribution of X only.

Let $\mathbf{V}, \mathbf{V}' \in \mathbb{SO}(d)$; observe that for f with $\|f\|_{BL} \leq 1$ given,

$$\begin{aligned}
&\Big| \big|\mathbb{E}_X f(X_{\mathbf{V}}^{(k)}) - \mathbb{E}f(\sigma Z)\big| - \big|\mathbb{E}_X f(X_{\mathbf{V}'}^{(k)}) - \mathbb{E}f(\sigma Z)\big| \Big| \\
&\leq \Big| \mathbb{E}_X f(X_{\mathbf{V}'}^{(k)}) - \mathbb{E}_X f(X_{\mathbf{V}}^{(k)})) \Big| \\
&= \Big| \mathbb{E}\Big[f\big(\langle X, \mathbf{V}_1' \rangle, \ldots, \langle X, \mathbf{V}_k' \rangle\big) - f\big(\langle X, \mathbf{V}_1 \rangle, \ldots, \langle X, \mathbf{V}_k \rangle\big) \Big| \mathbf{V}, \mathbf{V}' \Big] \Big| \\
&\leq \mathbb{E}\Big[\big| \big(\langle X, \mathbf{V}_1' - \mathbf{V}_1 \rangle, \ldots, \langle X, \mathbf{V}_k' - \mathbf{V}_k \rangle\big) \big| \Big| \mathbf{V}, \mathbf{V}' \Big] \\
&\leq \sqrt{ \sum_{j=1}^{k} |\mathbf{V}_j' - \mathbf{V}_j|^2 \mathbb{E} \left\langle X, \frac{\mathbf{V}_j' - \mathbf{V}_j}{|\mathbf{V}_j' - \mathbf{V}_j|} \right\rangle^2 } \\
&\leq \rho(\mathbf{V}, \mathbf{V}') \sqrt{B}.
\end{aligned}$$

It follows that

$$\begin{aligned}
&\Big| d_{BL}(X_{\mathbf{V}}^{(k)}, \sigma Z) - d_{BL}(X_{\mathbf{V}'}^{(k)}, \sigma Z) \Big| \\
&= \left| \sup_{\|f\|_{BL} \leq 1} \big|\mathbb{E}_X f(X_{\mathbf{V}}^{(k)}) - \mathbb{E}f(\sigma Z)\big| - \sup_{\|f\|_{BL} \leq 1} \big|\mathbb{E}_X f(X_{\mathbf{V}'}^{(k)}) - \mathbb{E}f(\sigma Z)\big| \right| \\
&\leq \sup_{\|f\|_{BL} \leq 1} \Big| \big|\mathbb{E}_X f(X_{\mathbf{V}}^{(k)}) - \mathbb{E}f(\sigma Z)\big| - \big|\mathbb{E}_X f(X_{\mathbf{V}'}^{(k)}) - \mathbb{E}f(\sigma Z)\big| \Big| \\
&\leq \rho(\mathbf{V}, \mathbf{V}') \sqrt{B},
\end{aligned}$$

thus $d_{BL}(X_{\mathbf{V}}^{(k)}, \sigma Z)$ is a Lipschitz function on $\mathbb{SO}(d)$, with Lipschitz constant $\sqrt{B}$. Applying the concentration of measure inequality in Corollary 5.17 thus implies that

$$\mathbb{P}\Big[|d_{BL}(X_{\mathbf{V}}^{(k)}, \sigma Z) - \mathbb{E}d_{BL}(X_{\mathbf{V}}^{(k)}, \sigma Z)| > \epsilon \Big] < C e^{-\frac{cd\epsilon^2}{B}}.$$

□

The final component of the proof of Theorem 6.7 is to estimate the so-called average distance to average.

Theorem 6.10 *With notation as in the previous theorems,*

$$\mathbb{E}d_{BL}(X_{\mathbf{V}}^{(k)}, \sigma Z) \leq C \left(\frac{k^{k-1} B^{k+2}}{d^2} \right)^{\frac{1}{3k+4}} + \frac{\sigma[\sqrt{k}(A+1)+k]}{d-1}.$$

Proof Let $\mathbf{V} \in \mathbb{SO}(d)$ and let $\mathbf{U}$ be a random orthogonal matrix, independent of X. Then the function

$$\mathbf{V} \longmapsto d_{BL}(X_{\mathbf{V}}^{(k)}, X_{\mathbf{U}}^{(k)})$$

can be viewed as a random variable, if $\mathbf{V}$ is now taken to be distributed according to Haar measure. The essential idea of the proof is to view this random variable as the supremum of a stochastic process: for $f : \mathbb{R}^k \to \mathbb{R}$ with $\|f\|_{BL} \leq 1$, let

$$X_f = X_f(\mathbf{V}) := \mathbb{E}_X f(X_{\mathbf{V}}^{(k)}) - \mathbb{E} f(X_{\mathbf{U}}^{(k)}),$$

where again $\mathbb{E}_X f(X_{\mathbf{V}}^{(k)})$ indicates expectation with respect to X only. Then $\{X_f\}_f$ is a centered stochastic process indexed by the set of functions f on $\mathbb{R}^k$ with $\|f\|_{BL} \leq 1$, and

$$d_{BL}(X_{\mathbf{V}}^{(k)}, X_{\mathbf{U}}^{(k)}) = \sup_{\|f\|_{BL} \leq 1} X_f.$$

Concentration of measure on the special orthogonal group implies that X_f is a sub-Gaussian process, as follows. Let $f : \mathbb{R}^k \to \mathbb{R}$ be Lipschitz with Lipschitz constant L and consider the function $G = G_f$ defined on $\mathbb{SO}(d)$ by

$$G(\mathbf{V}) := \mathbb{E}_X f(X_{\mathbf{V}}^{(k)}) = \mathbb{E}_X \left[f(\langle \mathbf{V}_1, X \rangle, \ldots, \langle \mathbf{V}_k, X \rangle) \right],$$

where $\mathbf{V}_i$ denotes the ith row of $\mathbf{V}$. It was shown in the course of the previous proof that $G(\mathbf{V})$ is a Lipschitz function on $\mathbb{SO}(d)$, with Lipschitz constant $L\sqrt{B}$, and so by Corollary 5.17 there are universal constants C, c such that

$$\mathbb{P}\left[|G(\mathbf{V}) - \mathbb{E}G(\mathbf{V})| > \epsilon \right] \leq Ce^{-\frac{cd\epsilon^2}{L^2 B}}. \tag{6.6}$$

It follows from Fubini's theorem that if $\mathbf{V}$ a random special orthogonal matrix, then $\mathbb{E}G(\mathbf{V}) = \mathbb{E}f(X_{\mathbf{U}}^{(k)})$, for $\mathbf{U}$ Haar-distributed and independent of X as above. Equation (6.6) can thus be restated as

$$\mathbb{P}\left[|X_f| > \epsilon \right] \leq C \exp\left[-\frac{cd\epsilon^2}{L^2 B} \right].$$

Note that $X_f - X_g = X_{f-g}$, and so

$$\mathbb{P}\left[\left| X_f - X_g \right| > \epsilon \right] \leq C \exp\left[\frac{-cd\epsilon^2}{2B|f - g|_L^2} \right] \leq C \exp\left[\frac{-cd\epsilon^2}{2B\|f - g\|_{BL}^2} \right].$$

The process $\{X_f\}$ therefore satisfies the sub-Gaussian increment condition for the metric space (BL_1, d^*), where BL_1 denotes the unit ball of the bounded-Lipschitz norm and $d^*(f, g) := \frac{\sqrt{B}}{\sqrt{cd}} \|f - g\|_{BL}$.

The idea is to apply Dudley's entropy bound to this sub-Gaussian process; however, it cannot be usefully applied to this infinite-dimensional indexing set. We therefore make several reductions, beginning with a truncation argument.

Let

$$\varphi_R(x) = \begin{cases} 1 & |x| \leq R, \\ R+1-|x| & R \leq |x| \leq R+1, \\ 0 & R+1 \leq |x|, \end{cases}$$

and define $f_R := f \cdot \varphi_R$; if $\|f\|_{BL} \leq 1$, then $\|f_R\|_{BL} \leq 2$. Since $|f(x) - f_R(x)| = 0$ if $x \in B_R$ and $|f(x) - f_R(x)| \leq 1$ for all $x \in \mathbb{R}^k$,

$$\left|\mathbb{E}_X f(X_{\mathbf{V}}^{(k)}) - \mathbb{E}_X f_R(X_{\mathbf{V}}^{(k)})\right| \leq \mathbb{P}_X\left[|X_{\mathbf{V}}^{(k)}| > R\right] \leq \frac{1}{R^2} \sum_{i=1}^{k} \mathbb{E}_X\left[\langle X, \mathbf{V}_i \rangle^2\right] \leq \frac{Bk}{R^2},$$

and the same holds if $\mathbb{E}_X$ is replaced by $\mathbb{E}$. It follows that $\left|X_f - X_{f_R}\right| \leq \frac{2Bk}{R^2}$. Consider therefore the process X_f indexed by $BL_{2,R+1}$ (with norm $\|\cdot\|_{BL}$), for some choice of R to be determined, where

$$BL_{2,R+1} := \left\{f : \mathbb{R}^k \to \mathbb{R} : \|f\|_{BL} \leq 2; f(x) = 0 \text{ if } |x| > R+1\right\};$$

what has been shown is that

$$\mathbb{E}\left[\sup_{\|f\|_{BL} \leq 1} X_f\right] \leq \mathbb{E}\left[\sup_{f \in BL_{2,R+1}} X_f\right] + \frac{2Bk}{R^2}. \tag{6.7}$$

The next step is to approximate functions in $BL_{2,R+1}$ by "piecewise linear" functions. Specifically, consider a cubic lattice of edge length ϵ in $\mathbb{R}^k$. Triangulate each cube of the lattice into simplices inductively as follows: in $\mathbb{R}^2$, add an extra vertex in the center of each square to divide the square into four triangles. To triangulate the cube of $\mathbb{R}^k$, first triangulate each facet as was described in the previous stage of the induction. Then add a new vertex at the center of the cube; connecting it to each of the vertices of each of the facets gives a triangulation into simplices. Observe that when this procedure is carried out, each new vertex added is on a cubic lattice of edge length $\frac{\epsilon}{2}$. Let $\mathcal{L}$ denote the supplemented lattice comprised of the original cubic lattice, together with the additional vertices needed for the triangulation. The number of sites of $\mathcal{L}$ within the ball of radius $R+1$ is then bounded by, e.g., $c\left(\frac{3R}{\epsilon}\right)^k \omega_k$, where ω_k is the volume of the unit ball in $\mathbb{R}^k$. It is classical that the volume ω_k of the unit ball in $\mathbb{R}^k$ is asymptotically given by $\frac{\sqrt{2}}{\sqrt{k\pi}}\left[\frac{2\pi e}{k}\right]^{\frac{k}{2}}$ as $k \to \infty$, so that the number of sites is bounded by $\frac{c}{\sqrt{k}}\left(\frac{c'R}{\epsilon\sqrt{k}}\right)^k$, for constants c, c' which are independent of d and k.

We now approximate $f \in BL_{2,R+1}$ by the function $\tilde{f}$ defined such that $\tilde{f}(x) = f(x)$ for $x \in \mathcal{L}$, and the graph of $\tilde{f}$ is determined by taking the convex hull of the vertices of the image under f of each k-dimensional simplex determined by $\mathcal{L}$. The resulting function $\tilde{f}$ still has $\|\tilde{f}\|_{BL} \leq 2$, and $\|f - \tilde{f}\|_\infty \leq \frac{\epsilon\sqrt{k}}{2}$, since the distance between points in the same simplex is bounded by $\frac{\epsilon\sqrt{k}}{2}$. Moreover, the function $\tilde{f}$ lies inside the finite-dimensional vector space of functions whose values are determined through the interpolation procedure described above by their values at the points of $\mathcal{L}$ within the ball of radius $R+1$. It thus follows that

$$\mathbb{E}\left[\sup_{f \in BL_{2,R+1}} X_f\right] \leq \mathbb{E}\left[\sup_{f \in BL_{2,R+1}} X_{\tilde{f}}\right] + \epsilon\sqrt{k}, \tag{6.8}$$

that the process $\{X_{\tilde{f}}\}_{f \in BL_{2,R+1}}$ is sub-Gaussian with respect to $\frac{\sqrt{B}}{\sqrt{cd}}\|\cdot\|_{BL}$, and that $\{\tilde{f} : f \in BL_{2,R+1}\}$ is the ball of radius 2 inside an M-dimensional normed space, with

$$M = \frac{c}{\sqrt{k}}\left(\frac{c'R}{\epsilon\sqrt{k}}\right)^k. \tag{6.9}$$

We have thus replaced a sub-Gaussian process indexed by a ball in an infinite-dimensional space with one indexed by a ball in a finite-dimensional space, where Dudley's bound can finally be applied. Let $T := \{\tilde{f} : f \in BL_{2,R+1}\}$; a classical volumetric argument gives that the covering numbers of the unit ball B of a finite-dimensional normed space $(X, \|\cdot\|)$ of dimension M can be bounded as $\mathcal{N}(B, \|\cdot\|, \epsilon) \leq \exp\left[M \log\left(\frac{3}{\epsilon}\right)\right]$. As a result,

$$\mathcal{N}\left(T, \frac{\sqrt{B}}{\sqrt{cd}}\|\cdot\|_{BL}, \epsilon\right) \leq \exp\left[M \log\left(\frac{c'\sqrt{B}}{\epsilon\sqrt{d}}\right)\right].$$

It follows from Dudley's entropy bound that

$$\mathbb{E}\sup_{f \in T} X_f \leq \int_0^{2\sqrt{\frac{B}{cd}}} \sqrt{M \log\left(\frac{c'\sqrt{B}}{\epsilon\sqrt{d}}\right)}\, d\epsilon = L\sqrt{\frac{MB}{d}}.$$

Combining this with (6.7) and (6.8) yields

$$\mathbb{E}\left[\sup_{\|f\|_{BL} \leq 1} \left(\mathbb{E}_X f(X_V) - \mathbb{E}f(X_U)\right)\right] \leq \frac{9kB}{R^2} + \epsilon\sqrt{k} + L\sqrt{\frac{MB}{d}}.$$

Using the value of M in terms of R given in (6.9) and choosing $\epsilon = \left(\frac{B(c'R)^k}{dk^{\frac{k+3}{2}}}\right)^{\frac{1}{k+2}}$ yields

$$\mathbb{E}\left[\sup_{\|f\|_{BL}\leq 1}(\mathbb{E}_X f(X_{\mathbf{V}}) - \mathbb{E}f(X_{\mathbf{U}}))\right] \leq \frac{9kB}{R^2} + \tilde{c}\left(\frac{BR^k}{d\sqrt{k}}\right)^{\frac{1}{k+2}}.$$

Finally, choosing $R = \left(dk^{\frac{2k+5}{2}}B^{k+1}\right)^{\frac{1}{3k+4}}$ yields

$$\mathbb{E}\left[\sup_{\|f\|_{BL}\leq 1}(\mathbb{E}_X f(X_{\mathbf{V}}) - \mathbb{E}f(X_{\mathbf{U}}))\right] \leq \tilde{L}\left(\frac{k^{k-1}B^{k+2}}{d^2}\right)^{\frac{1}{3k+4}}.$$

Combining this with Theorem 6.8 completes the proof. □

Notes and References

The Johnson–Lindenstrauss lemma was first proved in [61] as a step in showing that any mapping from an n-point set in a metric space X into ℓ_2 can be extended to a Lipschitz mapping of all of X into ℓ_2, with the Lipschitz constant at worst being multiplied by $c\sqrt{\log(n)}$. The lemma has found many applications in computer science and other areas; see the book [101] by Vempala for an extensive survey. The original proof used Gaussian random matrices; random orthogonal matrices were first used in [45], and there is now a large literature on alternative forms of randomness.

Dvoretzky's theorem first appeared in [40], and an enormous literature has grown up around the theorem and its applications. The recent book [4] has an extensive history and discussion of modern viewpoints and connections. The proof in the literature that is closest to the one given in Section 6.2 was given by Aubrun, Szarek, and Werner in [5], following an earlier approach of Schechtman [93].

Borel's theorem on the distribution of a coordinate of a random point on the sphere can be seen as a first example of the central limit theorem for convex bodies; that is, that marginals of the uniform measure on high-dimensional convex bodies are approximately Gaussian. Over the years, many authors made contributions on this problem; see in particular [15, 34, 43, 67, 99]. Finally, Klartag [68] showed that the typical *total variation distance* between a k-dimensional marginal of uniform measure on a convex body (or, more generally, any log-concave distribution) in $\mathbb{R}^d$ and the corresponding Gaussian distribution is small even when $k = d^{\epsilon}$ (for a specific universal constant $\epsilon \in (0, 1)$). See the recent book [14] for an extensive survey of the central limit theorem for convex bodies and related phenomena.

Theorem 6.7 first appeared in [76], and was an attempt to find the optimal dependence of k on d when the assumption of log-concavity was removed. The possible rate of growth of k as a function of d is indeed weaker than in the result of [68] for log-concave measures; k can grow only a bit more slowly than logarithmically with d, rather than polynomially. However, as the example following the theorem shows, either the log-concavity or some other additional assumption is necessary; with only the assumptions here, logarithmic-type growth of k in d is best possible for the bounded-Lipschitz metric.

7

Characteristic Polynomials and Connections to the Riemann ζ-function

In this chapter we give just a taste of the intriguing, and as yet unexplained, connection between zeros of the Riemann zeta function and eigenvalues of random matrices. A few pointers to the large literature on the subject are given in the end of chapter notes.

7.1 Two-Point Correlations and Montgomery's Conjecture

The Riemann zeta function is defined on complex numbers $s = \sigma + it$ with $\sigma > 1$ by either the Dirichlet series or the Euler product

$$\zeta(s) = \sum_{n=1}^{\infty} \frac{1}{n^s} = \prod_{p} \left(1 - \frac{1}{p^s}\right)^{-1},$$

where the product is over prime numbers p. The zeta function can be extended to the complex plane by analytic continuation and has a simple pole at $s = 1$ and trivial zeros at $s = -2n$ for $n = 1, 2, 3, \ldots$. It has been known since Riemann that the remaining zeros $\rho = \beta + i\gamma$ all lie in the "critical strip" $\{0 < \beta < 1\}$, and that the zeroes fall symmetrically about both the real line and the line $\beta = \frac{1}{2}$ (sometimes called the critical line), so that if ρ is a zero, so are $\overline{\rho}$, $1 - \rho$, and $1 - \overline{\rho}$. The Riemann Hypothesis (RH) is that all of the nontrivial zeros have $\beta = \frac{1}{2}$.

The interest in the Riemann Hypothesis lies in the connection between the zeros of the zeta function and the distribution of prime numbers. This is of course an enormous field of study, but the following gives a flavor of the connection. The von Mangoldt function $\Lambda(n)$ is defined by

$$\Lambda(n) = \begin{cases} \log p, & n = p^m, p \text{ prime}, m \geq 1; \\ 0, & \text{otherwise}, \end{cases}$$

and the Chebyshev function $\psi(x)$ is given by

$$\psi(x) = \sum_{n \leq x} \Lambda(n).$$

The Chebychev function is closely related to the prime counting function $\pi(x) = \{p \leq x : p \text{ prime}\}$, and the growth estimate

$$\psi(x) = x + O(x^{\frac{1}{2}}(\log x)^2)$$

is equivalent to RH.

Over the years, many approaches to proving RH have been explored. One is via spectral theory: let $\left\{\frac{1}{2} + it_n\right\}_{n \in \mathbb{N}}$ denote the nontrivial zeroes of the Riemann zeta function, so that RH is the statement that the t_n are all real. The Hilbert–Pólya conjecture is the statement that there is an unbounded Hermitian operator whose eigenvalues are $\{t_n\}_{n \in \mathbb{N}}$. This conjecture has inspired the idea that the eigenvalues of this mysterious operator may behave like the eigenvalues of a "random operator," which could be modeled by a (large) random matrix. Most classically, the model used was the GUE; i.e., a random matrix distributed according to Gaussian measure in Hermitian matrix space. The eigenvalues of random unitary matrices, and especially the spacings between them, have also been considered in this context, and in fact in many of the resulting conjectures, either random matrix model produces the same conjecture about the zeta zeroes. It turns out that indeed, when zeta zeros and eigenvalues of random matrices are compared numerically, their behavior matches astonishingly well. This connection was first noticed in a chance meeting of Hugh Montgomery and Freeman Dyson,[1] in the context of pair correlations.

In order to identify the appropriate correspondence between zeta zeroes and eigenvalues, a first important observation is that the local density of zeta zeroes at a given height within the critical strip is known. Let

$$n(T) := |\{\rho = \beta + i\gamma : \zeta(\rho) = 0, 0 < \gamma \leq T\}|,$$

$$n(T-) := |\{\rho = \beta + i\gamma : \zeta(\rho) = 0, 0 < \gamma < T\}|,$$

and

$$N(T) := \frac{n(T) + n(T-)}{2},$$

[1] Said encounter is now a much-loved anecdote illustrating, among other things, the value of department tea.

where in both $n(T)$ and $n(T-)$, the zeros are counted with multiplicity. The Riemann–von Mangoldt formula states that

$$N(T) = \frac{T}{2\pi} \log\left(\frac{T}{2\pi e}\right) + \frac{7}{8} + R(T) + S(T),$$

where $R(T) = O(\frac{1}{T})$ and

$$S(T) = \frac{1}{\pi} \arg\left(\zeta\left(\frac{1}{2} + iT\right)\right) = O(\log(T)).$$

In particular, this says that the local density of zeros at height T is approximately $\frac{1}{2\pi}\log\left(\frac{T}{2\pi}\right)$. Since the average density of the eigenvalues of $U \in \mathbb{U}(n)$ on the circle is $\frac{n}{2\pi}$, when comparing zeta zeroes and eigenvalues it makes sense to compare eigenvalues of $U \in \mathbb{U}(n)$ with zeros of ζ at height T, with

$$n = \log\left(\frac{T}{2\pi}\right),$$

so that the only natural parameter, namely the density of points, is the same.

We next transform the zeta zeros (and unitary eigenvalues correspondingly) so that the average spacing is 1. Suppose (in this case, really just for convenience of exposition) that RH holds, and order the zeros in the upper half-plane $\rho_n = \frac{1}{2} + it_n$ so that $0 < t_1 \le t_2 \le \cdots$. Define the "unfolded zeros" by

$$w_n := \frac{t_n}{2\pi} \log\left(\frac{t_n}{2\pi}\right),$$

so that by the Riemann–von Mangoldt formula,

$$\lim_{W\to\infty} \frac{1}{W}\left|\left\{w_n : 0 < w_n \le W\right\}\right| = 1.$$

For $\alpha < \beta$ and $W > 0$, define the functions

$$F_\zeta(\alpha, \beta; W) = \frac{1}{W}\left|\left\{w_j, w_k \in [0, W] : \alpha \le w_j - w_k < \beta\right\}\right|$$

and

$$F_\zeta(\alpha, \beta) = \lim_{W\to\infty} F_\zeta(\alpha, \beta; W).$$

That is, F_ζ gives the asymptotic density of pairs of zeta zeros separated by a prescribed distance.

The so-called **two-point correlation function** $R_{2,\zeta}(x)$ (we will see the connection with our earlier use of this term shortly) can be defined by the formula

$$F_\zeta(\alpha, \beta) = \int_\alpha^\beta R_{2,\zeta}(x)dx + \mathbb{1}_{[\alpha,\beta)}(0).$$

A fundamental goal in understanding the distribution of the zeros of ζ is to understand the behavior of the function $F_\zeta(\alpha,\beta)$, or equivalently, $R_{2,\zeta}(x)$. To do this, it is useful to generalize F_ζ as follows: given a test function f, let

$$F_{2,\zeta}(f;W) := \frac{1}{W}\sum_{0<w_j,w_k\le W} f(w_j - w_k).$$

In particular, $F_{2,\zeta}(\mathbb{1}_{[\alpha,\beta)};W) = F_\zeta(\alpha,\beta;W)$. For certain test functions, $\lim_{W\to\infty} F_{2,\zeta}(f;W)$ can be computed explicitly.

Theorem 7.1 (Montgomery [84]) *Let $f:\mathbb{R}\to\mathbb{R}$ be such that $\hat{f}(\xi) = \int_{-\infty}^{\infty} f(x)e^{2\pi i x\xi}\,dx$ is supported in $(-1,1)$. Then*

$$\lim_{W\to\infty} F_{2,\zeta}(f;W) = \int_{-\infty}^{\infty} f(x)\left[1-\left(\frac{\sin(\pi x)}{\pi x}\right)^2\right]dx.$$

Montgomery's theorem unfortunately does not apply to $f = \mathbb{1}_{[\alpha,\beta)}$; *Montgomery's conjecture* says that the theorem holds without the restriction on the support of $\hat{f}$, so that for $R_{2,\zeta}$ as defined above,

$$R_{2,\zeta}(x) = 1-\left(\frac{\sin(\pi x)}{\pi x}\right)^2.$$

Back on the random matrix side, let U be a random unitary matrix with eigenvalues $\{e^{i\theta_1},\ldots,e^{i\theta_n}\}$, and consider the rescaled eigenangles

$$\phi_j := \left(\frac{\theta_j}{2\pi}\right)n,$$

so that the average spacing is 1. Consider the analog to $F_{2,\zeta}(f;W)$ given by

$$F_{2,U}(f) := \frac{1}{n}\sum_{j,k=1}^{n} f(\phi_j - \phi_k).$$

Fix $\alpha<\beta$ and suppose that $0\notin[\alpha,\beta)$. Then

$$\begin{aligned}\mathbb{E}\left[F_{2,U}(\mathbb{1}_{[\alpha,\beta)})\right] &= \frac{1}{n}\mathbb{E}\left[\left|\{(j,k):\alpha\le\phi_j-\phi_k<\beta\}\right|\right]\\ &= \frac{1}{n}\mathbb{E}\left[\left|\left\{(j,k):\frac{2\pi\alpha}{n}+\theta_k\le\theta_j<\frac{2\pi\beta}{n}+\theta_k\right\}\right|\right].\end{aligned}$$

For $M \in \mathbb{N}$, divide $[0, 2\pi)$ into M ordered subintervals $I_1, I_2, \dots, I_M$ of equal length. It follows from the dominated convergence theorem that

$$\begin{aligned}
&\frac{1}{n}\mathbb{E}\left[\left|\left\{(j,k) : \frac{2\pi\alpha}{n} + \theta_k \leq \theta_j < \frac{2\pi\beta}{n} + \theta_k\right\}\right|\right] \\
&\quad = \frac{1}{n}\mathbb{E}\left[\lim_{M\to\infty} \sum_{\ell=1}^{M} \left|\left\{(j,k) : \theta_k \in I_\ell, \theta_j \in I_\ell + \left[\frac{2\pi\alpha}{n}, \frac{2\pi\beta}{n}\right)\right\}\right|\right] \\
&\quad = \frac{1}{n}\lim_{M\to\infty}\left(\sum_{\ell=1}^{M} \mathbb{E}\left[\mathcal{N}_{I_\ell}\mathcal{N}_{I_\ell + \left[\frac{2\pi\alpha}{n}, \frac{2\pi\beta}{n}\right)}\right]\right).
\end{aligned}$$

Now, since $0 \notin [\alpha, \beta)$, the intervals I_ℓ and $I_\ell + \left[\frac{2\pi\alpha}{n}, \frac{2\pi\beta}{n}\right)$ are disjoint when N is large enough, and so for ρ_2 the 2-point correlation function of the unitary eigenangle process (see Section 3.2),

$$\begin{aligned}
&\frac{1}{n}\lim_{M\to\infty}\left(\sum_{\ell=1}^{M} \mathbb{E}\left[\mathcal{N}_{I_\ell}\mathcal{N}_{I_\ell + \left[\frac{2\pi\alpha}{n}, \frac{2\pi\beta}{n}\right)}\right]\right) \\
&\qquad = \frac{1}{n}\lim_{M\to\infty}\left(\sum_{\ell=1}^{M} \frac{1}{(2\pi)^2}\int_{I_\ell}\int_{I_\ell + \left[\frac{2\pi\alpha}{n}, \frac{2\pi\beta}{n}\right)} \rho_2(x,y)dxdy\right) \\
&\qquad = \frac{1}{2\pi n}\int_{\frac{2\pi\alpha}{n}}^{\frac{2\pi\beta}{n}} \rho_2(0,y)dy \\
&\qquad = \frac{1}{2\pi}\int_{\alpha}^{\beta}\left[1 - \left(\frac{\sin(\pi u)}{n\sin\left(\frac{\pi u}{n}\right)}\right)^2\right]du,
\end{aligned}$$

using the form of $\rho_2(x, y)$ given by the chart on page 80.

If $0 \in [\alpha, \beta)$, then the argument above needs to be modified only slightly: in that case,

$$\sum_{\ell=1}^{M} \mathbb{E}\left[\mathcal{N}_{I_\ell}\mathcal{N}_{I_\ell + \left[\frac{2\pi\alpha}{n}, \frac{2\pi\beta}{n}\right)}\right] = \sum_{\ell=1}^{M} \mathbb{E}\left[\mathcal{N}_{I_\ell}^2 + \mathcal{N}_{I_\ell}\mathcal{N}_{\left(I_\ell + \left[\frac{2\pi\alpha}{n}, \frac{2\pi\beta}{n}\right)\right)\setminus I_\ell}\right].$$

We have seen (see the proof of Theorem 4.8) that $\mathrm{Var}(\mathcal{N}_{I_\ell}) = O\left(\left(\frac{n}{M}\right)^2\right)$, and so

$$\lim_{n\to\infty}\frac{1}{n}\lim_{M\to\infty}\sum_{\ell=1}^{M}\mathbb{E}[\mathcal{N}_{I_\ell}^2] = \lim_{n\to\infty}\frac{1}{n}\lim_{M\to\infty}\sum_{\ell=1}^{M}\mathbb{E}[\mathcal{N}_{I_\ell}] = 1,$$

whereas the second term can be treated exactly as in the previous case.

It thus follows that

$$\lim_{n\to\infty} \mathbb{E}\left[F_{2,U}(\mathbb{1}_{[\alpha,\beta)})\right] = \int_{\alpha}^{\beta} \left[1 - \left(\frac{\sin(\pi u)}{\pi u}\right)^2\right] du + \mathbb{1}_{[\alpha,\beta)}(0),$$

which exactly matches Montgomery's conjecture for the zeta zeros.

7.2 The Zeta Function and Characteristic Polynomials of Random Unitary Matrices

The computations in the previous section suggest that the eigenvalues of a random matrix, i.e., the zeros of its characteristic polynomial, behave similarly to the zeros of the Riemann zeta-function. An important next step was taken by Jon Keating and Nina Snaith in [65], who suggested that the value distribution of the characteristic polynomial of a random unitary matrix is a reasonable (local) model for the value distribution of the zeta-function itself. They demonstrated the validity of this idea by comparing new theorems in random matrix theory to known results and conjectures on the value distribution of the zeta-function; this in turn allowed them to formulate new conjectures on the value distribution of the zeta function, which are well supported by numerical work. This section gives a survey of some of the results and conjectures of Keating and Snaith, with the numerical evidence deferred until the next section.

The following theorem and conjecture are the two main points of comparison on the zeta side.

Theorem 7.2 (Selberg) *Let $E \subseteq \mathbb{C}$ be a rectangle. Then*

$$\lim_{T\to\infty} \frac{1}{T}\mu\left\{t : T \le t \le 2T, \frac{\log\zeta\left(\frac{1}{2}+it\right)}{\sqrt{\frac{1}{2}\log\log(\frac{T}{2\pi})}} \in E\right\} = \frac{1}{2\pi}\iint_E e^{-\frac{1}{2}(x^2+y^2)}dxdy,$$

where μ denotes Lebesgue measure on the line.

Conjecture 7.3 (Moment conjecture) *Let $\lambda \ge 0$. There is a function $f(\lambda)$ such that*

$$\lim_{T\to\infty} \frac{1}{(\log(T))^{\lambda^2}} \frac{1}{T}\int_0^T \left|\zeta\left(\frac{1}{2}+it\right)\right|^{2\lambda} dt = f(\lambda)a(\lambda),$$

where the arithmetic factor $a(\lambda)$ *is given by*

$$a(\lambda) = \prod_p \left[\left(1 - \frac{1}{p}\right)^{\lambda^2} \left(\sum_{m=0}^{\infty} \left(\frac{\Gamma(\lambda + m)}{m!\,\Gamma(\lambda)} \right)^2 \frac{1}{p^m} \right) \right].$$

(It is traditional to separate this arithmetic factor, which comes from number-theoretic considerations, rather than incorporating the unknown function $f(\lambda)$ into it, but of course one could simply state the conjecture as asserting the existence of the limit on the left-hand side.) The conjecture is trivially true when $\lambda = 0$, with $f(0) = 1$. It is known to be true when $\lambda = 1, 2$ with $f(1) = 1$ and $f(2) = \frac{1}{12}$. Aside from that, the conjecture is open, and prior to the work of Keating and Snaith, there were only even conjectured values for $f(\lambda)$ at $\lambda = 3, 4$.

On the random matrix side, consider the characteristic polynomial

$$Z(U, \theta) := \det(I - Ue^{-i\theta}) = \prod_{j=1}^{n} (1 - e^{i(\theta_j - \theta)}),$$

where $U \in \mathbb{U}(n)$ has eigenvalues $\{e^{i\theta_1}, \ldots, e^{i\theta_n}\}$. The following result is then the analog of Theorem 7.2.

Theorem 7.4 (Keating–Snaith) *Let* $Z(U,\theta) = \det(I - Ue^{-i\theta})$ *be the characteristic polynomial of a random matrix* $U \in \mathbb{U}(n)$, *and let* $E \subseteq \mathbb{C}$ *be a rectangle. Then*

$$\lim_{n\to\infty} \mathbb{P}\left[\frac{\log(Z(U,\theta))}{\sqrt{\frac{1}{2}\log(n)}} \in E \right] = \frac{1}{2\pi} \iint_E e^{-\frac{1}{2}(x^2+y^2)} dx dy.$$

In the case of Conjecture 7.3, the analogous limit on the random matrix side not only exists but can actually be computed, as follows.

Theorem 7.5 (Keating–Snaith) *Let* $Z(U,\theta) = \det(I - Ue^{-i\theta})$ *be the characteristic polynomial of a random matrix* $U \in \mathbb{U}(n)$, *and let* $\lambda \in \mathbb{C}$ *with* $\mathrm{Re}(\lambda) > -\frac{1}{2}$. *Then*

$$\lim_{n\to\infty} \frac{1}{n^{\lambda^2}} \mathbb{E}\left[|Z(U,\theta)|^{2\lambda} \right] = \frac{G^2(1+\lambda)}{G(1+2\lambda)},$$

where G *is the Barnes G-function, defined by*

$$G(1+z) = (2\pi)^{\frac{z}{2}} e^{-[(1+\gamma)z^2+z]/2} \prod_{j=1}^{\infty} \left[\left(1 + \frac{z}{j}\right)^j e^{-z+\frac{z^2}{2j}} \right],$$

with γ denoting the Euler–Mascheroni constant. In particular, for $k \in \mathbb{N}$,

$$\lim_{n\to\infty} \frac{1}{n^{k^2}} \mathbb{E}\left[|Z(U,\theta)|^{2k}\right] = \prod_{j=0}^{k-1} \frac{j!}{(j+k)!}.$$

This result had a huge impact because it suggested the following conjecture for the value of the function $f(\lambda)$ in the moment conjecture.

Conjecture 7.6 (Keating–Snaith) *Let $f(\lambda)$ be as in the moment conjecture, and let*

$$f_{\mathbb{U}}(\lambda) := \frac{G^2(1+\lambda)}{G(1+2\lambda)} = \lim_{n\to\infty} \frac{1}{n^{\lambda^2}} \mathbb{E}\left[|Z(U,\theta)|^{2\lambda}\right].$$

Then for $\mathrm{Re}(\lambda) > -\frac{1}{2}$,

$$f(\lambda) = f_{\mathbb{U}}(\lambda).$$

As mentioned above, the values of f were only known previously at $\lambda = 1, 2$. Aside from that, there were conjectures, by Conrey and Ghosh at $\lambda = 3$ and Conrey and Gonek at $\lambda = 4$, which match the Keating–Snaith conjecture above.

The key ingredient for the results on the characteristic polynomial stated above is the following explicit expression for the moment generating function of $\log(Z(U,\theta))$.

Lemma 7.7 *Let $U \in \mathbb{U}(n)$ be a random unitary matrix, and let $Z(U,\theta) = \det(I - Ue^{-i\theta})$ denote the characteristic polynomial of U. Let $s,t \in \mathbb{C}$ with* $\mathrm{Re}(t \pm s) > -2$. *Then for all $\theta \in [0, 2\pi)$,*

$$\mathbb{E}\left[|Z(U,\theta)|^t e^{is\,\mathrm{Im}\log(Z(U,\theta))}\right] = \prod_{k=1}^{n} \frac{\Gamma(k)\Gamma(k+t)}{\Gamma\left(k+\frac{t}{2}-\frac{s}{2}\right)\Gamma\left(k+\frac{t}{2}+\frac{s}{2}\right)}. \tag{7.1}$$

Proof First note that since the distribution of the eigenvalues of U is rotationally invariant,

$$Z(U,\theta) \stackrel{d}{=} Z(U,0)$$

for all θ; we will therefore immediately specialize to $\theta = 0$, and write $Z := Z(U,0)$.

Note also that, with probability one, none of the eigenangles θ_j are 0 and so $1 - e^{i\theta_j}$ is in the right half-plane where we may unambiguously set the argument in $(-\pi, \pi)$.

By the Weyl integration formula,

$$
\begin{aligned}
&\mathbb{E}\left[|Z|^t e^{is\,\mathrm{Im}\log(Z)}\right] \\
&= \mathbb{E}\left[\left|\prod_{j=1}^{n}(1-e^{i\theta_j})\right|^t e^{is\sum_{j=1}^{n}\mathrm{Im}(\log(1-e^{i\theta_j}))}\right] \\
&= \frac{1}{(2\pi)^n n!}\int_0^{2\pi}\cdots\int_0^{2\pi}\left|\prod_{j=1}^{n}(1-e^{i\theta_j})\right|^t e^{-is\sum_{j=1}^{n}\sum_{m=1}^{\infty}\frac{\sin(m\theta_j)}{m}} \\
&\qquad\qquad \times \prod_{1\le j<k\le n}|e^{i\theta_j}-e^{i\theta_k}|^2 d\theta_1\cdots d\theta_n,
\end{aligned}
\tag{7.2}
$$

where we have also expanded the logarithm in the exponent.

This integral can (with some effort) be evaluated using Selberg's integral formula, stated as follows.

For $a, b, \alpha, \beta, \delta \in \mathbb{C}$ with $\mathrm{Re}(a), \mathrm{Re}(b), \mathrm{Re}(\alpha), \mathrm{Re}(\beta) > 0$; $\mathrm{Re}(\alpha+\beta) > 1$; and

$$
-\frac{1}{n} < \mathrm{Re}(\delta) < \min\left\{\frac{\mathrm{Re}(\alpha)}{n-1}, \frac{\mathrm{Re}(\beta)}{n-1}, \frac{\mathrm{Re}(\alpha+\beta+1)}{2(n-1)}\right\},
$$

$$
\begin{aligned}
&J(a,b,\alpha,\beta,\delta,n) \\
&:= \int_{-\infty}^{\infty}\cdots\int_{-\infty}^{\infty}\left|\prod_{1\le j<k\le n}(x_j-x_k)\right|^{2\delta}\prod_{j=1}^{n}(a+ix_j)^{-\alpha}(b-ix_j)^{-\beta}dx_1\cdots dx_n \\
&= \frac{(2\pi)^n}{(a+b)^{(\alpha+\beta)n-\delta n(n-1)-n}}\prod_{j=0}^{n-1}\frac{\Gamma(1+\delta+j\delta)\Gamma(\alpha+\beta-(n+j-1)\delta-1)}{\Gamma(1+\delta)\Gamma(\alpha-j\delta)\Gamma(\beta-j\delta)}.
\end{aligned}
$$

Examining the factors in the integrand in (7.2),

$$
\begin{aligned}
\prod_{1\le j<k\le n}|e^{i\theta_j}-e^{i\theta_k}|^2 &= \prod_{1\le j<k\le n}\left|e^{\frac{i(\theta_j-\theta_k)}{2}}-e^{-\frac{i(\theta_j-\theta_k)}{2}}\right|^2 \\
&= 2^{n(n-1)}\prod_{1\le j<k\le n}\left|\sin\left(\tfrac{\theta_j}{2}-\tfrac{\theta_k}{2}\right)\right|^2,
\end{aligned}
$$

and similarly,

$$
\left|\prod_{j=1}^{n}(1-e^{i\theta_j})\right| = 2^n\left|\prod_{j=1}^{n}\sin\left(\tfrac{\theta_j}{2}\right)\right|.
$$

For each $\theta_j \in (0, 2\pi)$,

$$\sum_{m=1}^{\infty} \frac{\sin(m\theta_j)}{m} = \frac{\pi - \theta_j}{2}$$

(this is just the expansion of $\frac{\pi - x}{2}$ as a Fourier series on $(0, 2\pi)$), and so

$$\mathbb{E}\left[|Z|^t e^{is \operatorname{Im}\log(Z)}\right]$$
$$= \frac{2^{n(n-1)+tn}}{(2\pi)^n n!} \int_0^{2\pi} \cdots \int_0^{2\pi} \left| \prod_{j=1}^n \sin\left(\frac{\theta_j}{2}\right) \right|^t \prod_{j=1}^n e^{-\frac{is(\pi-\theta_j)}{2}}$$
$$\times \prod_{1 \le j < k \le n} \left| \sin\left(\frac{\theta_j}{2} - \frac{\theta_k}{2}\right) \right|^2 d\theta_1 \cdots d\theta_n,$$

Letting $\phi_j = \frac{\theta_j - \pi}{2}$ now gives that

$$\mathbb{E}\left[|Z|^t e^{is \operatorname{Im}\log(Z)}\right] = \frac{2^{n^2+tn}}{(2\pi)^n n!} \int_{-\frac{\pi}{2}}^{\frac{\pi}{2}} \cdots \int_{-\frac{\pi}{2}}^{\frac{\pi}{2}} \left| \prod_{j=1}^n \cos(\phi_j) \right|^t \prod_{j=1}^n e^{is\phi_j}$$
$$\times \prod_{1 \le j < k \le n} \left| \sin(\phi_j - \phi_k) \right|^2 d\phi_1 \cdots d\phi_n$$
$$= \frac{2^{n^2+tn}}{(2\pi)^n n!} \int_{-\frac{\pi}{2}}^{\frac{\pi}{2}} \cdots \int_{-\frac{\pi}{2}}^{\frac{\pi}{2}} \left| \prod_{j=1}^n \cos(\phi_j) \right|^t \prod_{j=1}^n e^{is\phi_j}$$
$$\times \prod_{1 \le j < k \le n} \left| \sin(\phi_j)\cos(\phi_k) - \cos(\phi_j)\sin(\phi_k) \right|^2 d\phi_1 \cdots d\phi_n$$
$$= \frac{2^{n^2+tn}}{(2\pi)^n n!} \int_{-\frac{\pi}{2}}^{\frac{\pi}{2}} \cdots \int_{-\frac{\pi}{2}}^{\frac{\pi}{2}} \left| \prod_{j=1}^n \cos(\phi_j) \right|^{t+2(n-1)} \prod_{j=1}^n e^{is\phi_j}$$
$$\times \prod_{1 \le j < k \le n} \left| \tan(\phi_j) - \tan(\phi_k) \right|^2 d\phi_1 \cdots d\phi_n.$$

Now letting $x_j = \tan(\phi_j)$ so that

$$\cos(\phi_j) = \frac{1}{\sqrt{1 + x_j^2}} \qquad \sin(\phi_j) = \frac{x_j}{\sqrt{1 + x_j^2}}$$

gives that

$$\mathbb{E}\left[|Z|^t e^{is \operatorname{Im} \log(Z)}\right]$$

$$= \frac{2^{n^2+tn}}{(2\pi)^n n!} \int_{-\infty}^{\infty} \cdots \int_{-\infty}^{\infty} \prod_{j=1}^{n} \left[\left(\frac{1}{\sqrt{1+x_j^2}}\right)^{t+2n} \left(\frac{1+ix_j}{\sqrt{1+x_j^2}}\right)^{s}\right]$$

$$\times \prod_{1\le j<k\le n} |x_j - x_k|^2 dx_1 \cdots dx_n$$

$$= \frac{2^{n^2+tn}}{(2\pi)^n n!} \int_{-\infty}^{\infty} \cdots \int_{-\infty}^{\infty} \prod_{j=1}^{n} \left[(1+ix_j)^{-(n+\frac{t}{2}-\frac{s}{2})}(1-ix_j)^{-(n+\frac{t}{2}+\frac{s}{2})}\right]$$

$$\times \prod_{1\le j<k\le n} |x_j - x_k|^2 dx_1 \cdots dx_n$$

$$= \frac{2^{n^2+tn}}{(2\pi)^n n!} J\left(1,1,n+\frac{t}{2}-\frac{s}{2}, n+\frac{t}{2}+\frac{s}{2}, 1, n\right).$$

The conditions for Selberg's integral formula to apply are that $\operatorname{Re}(t \pm s) > -2$; under these conditions, the formula yields

$$\mathbb{E}\left[|Z|^t e^{is \operatorname{Im} \log(Z)}\right] = \prod_{k=1}^{n} \frac{\Gamma(k)\Gamma(k+t)}{\Gamma\left(k+\frac{t}{2}-\frac{s}{2}\right)\Gamma\left(k+\frac{t}{2}+\frac{s}{2}\right)}$$

as claimed. □

Proof of Theorem 7.4 As before, it suffices to consider $Z = Z(U,0)$. For $t, s \in \mathbb{R}$, let

$$q = \frac{s}{\sqrt{\frac{1}{2}\log(n)}} \qquad r = \frac{t}{\sqrt{\frac{1}{2}\log(n)}}.$$

The moment generating function of the complex random variable $\frac{\log(Z)}{\sqrt{\frac{1}{2}\log(n)}}$ is

$$M(t,s) = \mathbb{E}\left[|Z|^r e^{iq \operatorname{Im}(\log(Z))}\right] = \prod_{k=1}^{n} \frac{\Gamma(k)\Gamma(k+r)}{\Gamma\left(k+\frac{r}{2}-\frac{q}{2}\right)\Gamma\left(k+\frac{r}{2}+\frac{q}{2}\right)},$$

and so

$$\log(M(t,s)) = \sum_{k=1}^{n} \left[\log(\Gamma(k)) + \log(\Gamma(k+r)) - \log\left(\Gamma\left(k+\frac{r}{2}-\frac{q}{2}\right)\right) - \log\left(\Gamma\left(k+\frac{r}{2}+\frac{q}{2}\right)\right)\right]. \tag{7.3}$$

The idea is to evaluate the limit in a neighborhood of $(t,s) = (0,0)$ by first expanding this expression as a power series in s and t and then taking the limit as $n \to \infty$; the claimed central limit theorem follows if we can show that for s and t small enough,

$$\lim_{n\to\infty} \log(M(t,s)) = t^2 - s^2 = \log \mathbb{E}[e^{tZ_1+isZ_2}],$$

where Z_1 and Z_2 are independent standard Gaussian random variables.

Now, by definition, $M(0,0) = 1$ and so $\log(M(0,0)) = 0$.

For $\ell \geq 0$, let

$$\psi^{(\ell)}(z) = \frac{d^{\ell+1}[\log(\Gamma(z))]}{dz^\ell} = (-1)^{\ell+1} \int_0^\infty \frac{t^\ell e^{-zt}}{1-e^{-t}} dt \tag{7.4}$$

denote the ℓth polygamma function. Then by (7.3),

$$\frac{\partial}{\partial s}[\log(M(t,s))] = \frac{1}{\sqrt{2\log(n)}} \sum_{k=1}^n \left[\psi^{(0)}\left(k + \frac{t-s}{\sqrt{2\log(n)}}\right) - \psi^{(0)}\left(k + \frac{t+s}{\sqrt{2\log(n)}}\right)\right],$$

and so

$$\left.\frac{\partial}{\partial s}[\log(M(t,s))]\right|_{(0,0)} = 0.$$

Similarly,

$$\left.\frac{\partial}{\partial t}[\log(M(t,s))]\right|_{(0,0)} = 0 \qquad \text{and} \qquad \left.\frac{\partial^2}{\partial t \partial s}[\log(M(t,s))]\right|_{(0,0)} = 0,$$

and so the power series expansion of $\log(M(t,s))$ has no terms of order 0 or 1 and no st term.

Continuing,

$$\frac{\partial^2}{\partial s^2}[\log(M(t,s))] = -\frac{1}{2\log(n)} \sum_{k=1}^n \left[\psi^{(1)}\left(k + \frac{t-s}{\sqrt{2\log(n)}}\right) + \psi^{(1)}\left(k + \frac{t+s}{\sqrt{2\log(n)}}\right)\right],$$

and so

$$\left.\frac{\partial^2}{\partial s^2}[\log(M(t,s))]\right|_{(0,0)} = -\frac{1}{\log(n)} \sum_{k=1}^n \psi^{(1)}(k).$$

Using the integral expression for $\psi^{(1)}$ from (7.4),

$$\begin{aligned}
\sum_{k=1}^{n} \psi^{(1)}(k) &= \int_0^\infty \frac{t}{1-e^{-t}} \left(\sum_{k=1}^{n} e^{-kt} \right) dt \\
&= \int_0^\infty \frac{te^{-t}(1-e^{-nt})}{(1-e^{-t})^2} dt \\
&= \int_0^\infty [(t(1-e^{-nt})] \frac{d}{dt} \left(\frac{e^{-t}}{1-e^{-t}} \right) dt \\
&= \int_0^\infty \left(\frac{e^{-t}}{1-e^{-t}} \right) \left[(1-e^{-nt}) + nte^{-nt} \right] dt.
\end{aligned}$$

Rearranging and re-expanding the geometric sums, this last expression is

$$\begin{aligned}
&\int_0^\infty \left[\frac{e^{-t}(1-e^{-nt})}{1-e^{-t}} + \frac{nte^{-(n+1)t}}{1-e^{-t}} \right] dt \\
&= \int_0^\infty \left[\left(\sum_{k=1}^{n} e^{-kt} \right) + nt \left(\sum_{k=n+1}^{\infty} e^{-kt} \right) \right] dt \\
&= \sum_{k=1}^{n} \frac{1}{k} + n \sum_{k=n+1}^{\infty} \frac{1}{k^2} = \log(n) + O(1).
\end{aligned}$$

It thus follows that

$$\lim_{n\to\infty} \left(\frac{s^2}{2} \frac{\partial^2}{\partial s^2} [\log(M(t,s))] \bigg|_{(0,0)} \right) = -\frac{s^2}{2}.$$

The proof that

$$\lim_{n\to\infty} \left(\frac{t^2}{2} \frac{\partial^2}{\partial t^2} [\log(M(t,s))] \bigg|_{(0,0)} \right) = \frac{t^2}{2}$$

is nearly identical. It thus remains to show that in some neighborhood of $(0,0)$, for s and t fixed,

$$\lim_{n\to\infty} \left(\sum_{j+\ell\geq 3} \frac{s^j t^\ell}{j!\,\ell!} \frac{\partial^{j+\ell}}{\partial s^j \partial t^\ell} [\log(M(t,s))] \bigg|_{(0,0)} \right) = 0.$$

From (7.3),

$$\frac{\partial^{j+\ell}}{\partial s^j \partial t^\ell}[\log(M(t,s))]\Big|_{(0,0)} = \sum_{k=1}^{n}\Bigg[\mathbb{1}_{j=0}\psi^{\ell-1}(k)\left(\frac{2}{\log(n)}\right)^{\frac{\ell}{2}} - \psi^{(j+\ell-1)}(k)\left(\frac{1+(-1)^j}{(2\log(n))^{\frac{j+\ell}{2}}}\right)\Bigg].$$

We will just consider the second term; the first is essentially the same, but slightly easier.

Using the integral representation of the polygamma function and integration by parts as above,

$$\begin{aligned}
\sum_{k=1}^{n}\psi^{(j+\ell-1)}(k) &= (-1)^{j+\ell}\sum_{k=1}^{n}\left[\int_0^\infty \frac{t^{j+\ell-1}e^{-kt}}{1-e^{-t}}dt\right] \\
&= (-1)^{j+\ell}\int_0^\infty \frac{t^{j+\ell-1}e^{-t}[1-e^{-nt}]}{(1-e^{-t})^2}dt \\
&= (-1)^{j+\ell}\int_0^\infty \left(\frac{e^{-t}}{1-e^{-t}}\right)\Big[(j+\ell-1)t^{j+\ell-2} \\
&\qquad -(j+\ell-1)t^{j+\ell-2}e^{-nt} + nt^{j+\ell-1}e^{-nt}\Big]dt.
\end{aligned} \tag{7.5}$$

Now, for $n \geq 2$,

$$\int_0^\infty \frac{t^{n-1}}{e^t-1}dt = \Gamma(n)\zeta(n),$$

and so

$$\begin{aligned}
&\int_0^\infty \left(\frac{e^{-t}}{1-e^{-t}}\right)(j+\ell-1)t^{j+\ell-2}dt \\
&= (j+\ell-1)\int_0^\infty \frac{t^{j+\ell-2}}{e^t-1}dt = (j+\ell-1)!\,\zeta(j+\ell-1).
\end{aligned}$$

For $j+\ell \geq 3$, this last expression is bounded by $(j+\ell)!\,\zeta(2)$, and so it follows that

$$\begin{aligned}
&\left|\sum_{j+\ell\geq 3}\frac{s^j t^\ell(-1)^{j+\ell}(1-(-1)^j)}{j!\,\ell!\,(2\log(n))^{\frac{j+\ell}{2}}}\int_0^\infty\left(\frac{e^{-t}}{1-e^{-t}}\right)(j+\ell-1)t^{j+\ell-2}dt\right| \\
&\leq 2\zeta(2)\sum_{m=3}^{\infty}\sum_{j=0}^{m}\left[\binom{m}{j}|s|^j|t|^{m-j}\left(\sqrt{\frac{1}{2\log(n)}}\right)^m\right] \\
&\leq \frac{2\zeta(2)}{(2\log(n))^{\frac{3}{2}}}\sum_{m=3}^{\infty}(|s|+|t|)^m,
\end{aligned}$$

which tends to zero as $n \to \infty$, as long as $|s| + |t| < 1$.

Considering the absolute value of the remaining terms in (7.5),

$$\int_0^\infty \left(\frac{e^{-t}}{1-e^{-t}}\right)\left[(j+\ell-1)t^{j+\ell-2} + nt^{j+\ell-1}\right]e^{-nt}dt$$
$$= \int_0^\infty \frac{\left[(j+\ell-1)t^{j+\ell-2} + nt^{j+\ell-1}\right]e^{-nt}}{e^t - 1}dt$$
$$\leq \int_0^\infty \left[(j+\ell-1)t^{j+\ell-3} + nt^{j+\ell-2}\right]e^{-nt}dt$$
$$= \frac{1}{n^{j+\ell-2}}\int_0^\infty [(j+\ell-1)y^{j+\ell-3} + y^{j+\ell-2}]e^{-y}dy$$
$$= \frac{1}{n^{j+\ell-2}}[(j+\ell-1)\Gamma(j+\ell-2) + \Gamma(j+\ell-1)]$$
$$\leq \frac{2(j+\ell-1)!}{n^{j+\ell-2}},$$

where we have used the estimate $e^t - 1 \geq t$ in the first inequality.

We can now bound the contribution to the tail of the power series for $\log(M(t,s))$:

$$\sum_{j+\ell\geq 3} \frac{2|s|^j|t|^\ell}{j!\,\ell!\,(2\log(n))^{\frac{j+\ell}{2}}}\int_0^\infty \left(\frac{e^{-t}}{1-e^{-t}}\right)\left[(j+\ell-1)t^{j+\ell-2} + nt^{j+\ell-1}\right]e^{-nt}dt$$
$$\leq \frac{4}{n(2\log(n))^{\frac{3}{2}}}\sum_{m=3}^{\infty}\sum_{j=0}^{m}\binom{m}{j}|s|^j|t|^{m-j}$$
$$= \frac{4}{n(2\log(n))^{\frac{3}{2}}}\sum_{m=3}^{\infty}(|s|+|t|)^m,$$

which again tends to zero as long as $|s| + |t| < 1$.

We have therefore shown that if $|s| + |t| < 1$, then

$$\lim_{n\to\infty}\log(M(t,s)) = \frac{t^2}{2} - \frac{s^2}{2},$$

which completes the proof of the bivariate Gaussian limit. □

Proof of Theorem 7.5 As usual, set $\theta = 0$ and consider $Z = Z(U, 0)$. Letting $s = 0$ and $t = 2\lambda$ in the statement of Lemma 7.7 gives that

$$\frac{1}{n^{\lambda^2}}\mathbb{E}\left[|Z|^{2\lambda}\right] = \frac{1}{n^{\lambda^2}}\prod_{k=1}^{n}\frac{\Gamma(k)\Gamma(k+2\lambda)}{(\Gamma(k+\lambda))^2}. \tag{7.6}$$

The Barnes G-function satisfies the functional equation $G(z+1) = \Gamma(z)G(z)$, so that

$$\Gamma(z) = \frac{G(z+1)}{G(z)}.$$

Expressing all the gamma functions in (7.6) as ratios of G functions in this way leads to massive cancellation, so that

$$\frac{1}{n^{\lambda^2}}\mathbb{E}\left[|Z|^{2\lambda}\right] = \frac{G^2(1+\lambda)}{G(1+2\lambda)}\left[\frac{G(n+1)G(n+1+2\lambda)}{n^{\lambda^2}G^2(n+1+\lambda)}\right],$$

using the fact that $G(1) = 1$.

To prove the theorem, it therefore suffices to show that for λ fixed,

$$\lim_{n\to\infty}\left[\frac{G(n+1)G(n+1+2\lambda)}{n^{\lambda^2}G^2(n+1+\lambda)}\right] = 1.$$

For z in the right half-plane with $|z|$ large, there is the following expansion of $\log(G(z+1))$:

$$\log(G(z+1)) = C + \frac{z}{2}\log(2\pi) + \left(\frac{z^2}{2} - \frac{1}{12}\right)\log(z) - \frac{3z^2}{4} + O\left(\frac{1}{z^2}\right).$$

Taking the logarithm of the expression above and using this expansion yields

$$\begin{aligned}
&\log\left[\frac{G(n+1)G(n+1+2\lambda)}{n^{\lambda^2}G^2(n+1+\lambda)}\right] \\
&= -\lambda^2\log(n) + \log(G(n+1)) + \log(G(n+2\lambda+1)) - 2\log(G(n+\lambda+1)) \\
&= -\lambda^2\log(n) + \left(\frac{n^2}{2} - \frac{1}{12}\right)\log(n) + \left(\frac{(n+2\lambda)^2}{2} - \frac{1}{12}\right)\log\left(n\left(1+\frac{2\lambda}{n}\right)\right) \\
&\quad - 2\left(\frac{(n+\lambda)^2}{2} - \frac{1}{12}\right)\log\left(n\left(1+\frac{\lambda}{n}\right)\right) - \frac{3n^2}{4} - \frac{3(n+2\lambda)^2}{4} \\
&\quad + \frac{3(n+\lambda)^2}{2} + O\left(\frac{1}{n^2}\right) \\
&= \left(\frac{n^2}{2} + 2n\lambda + 2\lambda^2 - \frac{1}{12}\right)\log\left(1+\frac{2\lambda}{n}\right) \\
&\quad - \left(n^2 + 2n\lambda + \lambda^2 - \frac{1}{6}\right)\log\left(1+\frac{\lambda}{n}\right) - \frac{3\lambda^2}{2} + O\left(\frac{1}{n^2}\right)
\end{aligned}$$

$$= \left(\frac{n^2}{2} + 2n\lambda + 2\lambda^2 - \frac{1}{12}\right)\left(\frac{2\lambda}{n} - \frac{2\lambda^2}{n^2} + O\left(\frac{1}{n^3}\right)\right)$$
$$- \left(n^2 + 2n\lambda + \lambda^2 - \frac{1}{6}\right)\left(\frac{\lambda}{n} - \frac{\lambda^2}{2n^2} + O\left(\frac{1}{n^3}\right)\right) - \frac{3\lambda^2}{2} + O\left(\frac{1}{n^2}\right)$$
$$= O\left(\frac{1}{n}\right),$$

where the implicit constants depend on λ. It thus follows that

$$\lim_{n\to\infty} \log\left[\frac{G(n+1)G(n+1+2\lambda)}{n^{\lambda^2}G^2(n+1+\lambda)}\right] = 0,$$

which completes the proof. □

7.3 Numerical and Statistical Work

While the connection between the zeta zeroes and the eigenvalues of random unitary matrices remains unexplained, there is ample numerical evidence. This section is mainly to advertise the existence of serious numerical work with some impressive pictures; for the reader looking for more involved numerical and statistical analysis, see the references in the end-of-section notes.

The earliest numerical work in this area was by Andrew Odlyzko, who generated large tables (and then much larger tables) of zeta zeroes, which he first used to investigate Montgomery's conjecture. Consider only the zeroes ρ with $\mathrm{Im}(\rho) > 0$, and order them (with multiplicity) by imaginary part: $0 < \mathrm{Im}(\rho_1) \leq \mathrm{Im}(\rho_2) \leq \cdots$. Odlyzko computed a large number of such zeros and found them all to be simple and lie on the critical line, so that they may be written

$$\rho_n = \frac{1}{2} + it_n.$$

Recall that Montgomery's conjecture dealt with the unfolded zeroes

$$w_n = \frac{t_n}{2\pi}\log\left(\frac{t_n}{2\pi}\right).$$

Rather than working with the unfolded zeroes, Odlyzko considered normalizing the spacings: for $n \in \mathbb{N}$,

$$\delta_n := \frac{t_{n+1} - t_n}{2\pi}\log\left(\frac{t_n}{2\pi}\right).$$

These two transformation are not meaningfully different for testing Montgomery's conjecture: either makes the average spacing equal to 1 at large height.

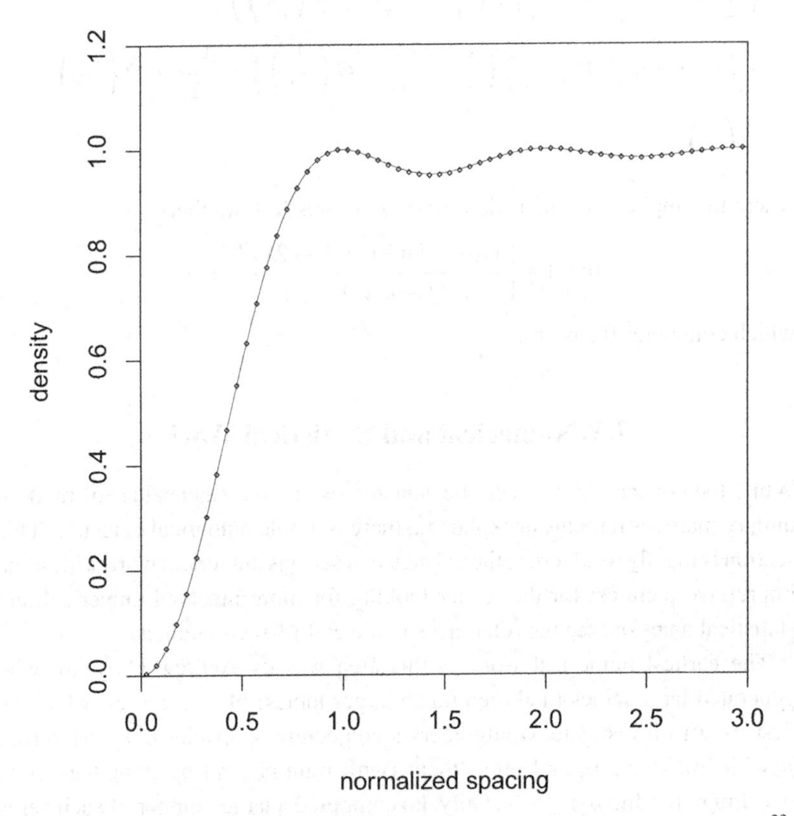

Figure 7.1 Pair correlations of (spacing-normalized) zeta zeroes around height 10^{23}, and the random matrix prediction. *Figure courtesy of Andrew Odlyzko.*

Figure 7.1 below illustrates the pair correlations for the data set

$$\{t_n : N+1 \leq n \leq N+M\},$$

where $N = 100000000000018097784511$ and $M = 203401872$. The interval $[0,3)$ is broken into subintervals $[\alpha, \beta)$ of length $\frac{1}{20}$, and for each interval $[\alpha, \beta)$,

$$a_{\alpha,\beta} := \frac{20}{M} \left|\{(n,k) : N+1 \leq n \leq N+M, k \geq 0, \delta_n + \cdots + \delta_{n+k} \in [\alpha, \beta)\}\right|.$$

According to Montgomery's conjecture, we should have

$$a_{\alpha,\beta} \approx 1 - \left(\frac{\sin(\pi\gamma)}{\pi\gamma}\right)^2,$$

for any $\gamma \in [\alpha, \beta)$. In the figure, each interval $[\alpha, \beta)$ has a point plotted at $x = \frac{\alpha+\beta}{2}$, $y = a_{\alpha,\beta}$; the solid line is the graph of $y = 1 - \left(\frac{\sin(\pi x)}{\pi x}\right)^2$.

A related connection also tested by Odlyzko is the distribution of nearest neighbor spacings. One can compute the predicted distribution of the δ_n based on the limiting pair correlation; the predicted distribution is the so-called Gaudin–Mehta distribution, which has a known, albeit not very explicit, density. Figure 7.2 below shows the empirical distribution of the δ_n for about 10^9 zeroes of the zeta function, at height approximately 10^{23}. Each plotted point has x-coordinate at the midpoint of an interval of length $\frac{1}{20}$ and height given by the proportion of the computed δ_n lying in that interval. The smooth curve is the density of the predicted distribution.

Statistics related to the normalized spacings are all about the local behavior of the zeta zeroes. More global features of Odlyzko's data were subjected to extensive statistical testing by Coram and Diaconis in [24]; below, we present two of their main results, illustrated with the relevant figures from [24].

Because global statistics of random matrices (e.g., the trace) depend on the entire ensemble of eigenvalues, a comparison with some corresponding feature of the zeta data requires one to first put blocks of zeta zeroes onto the unit circle. In their statistical analysis, Coram and Diaconis used a data set consisting of 50,000 consecutive zeros starting around the $10^{20\text{th}}$ zero, which is roughly at height $T = 1.5 \times 10^{19}$, corresponding to matrices of size $n = 42$. First the data set was divided into blocks of length 43. Each block was then wrapped around the unit circle so that the first and last zero were both at $z = 1$, and then the circle was randomly rotated. More concretely, for each block $\left\{\frac{1}{2} + it_n, \ldots, \frac{1}{2} + it_{n+43}\right\}$, let $\delta_j = t_{n+j} - t_{n+j-1}$ for $j = 1, \ldots, 42$, and let $\Delta_j := \sum_{k=1}^{j} \delta_k$. Then the wrapped zeta zeroes are the (random) points $X_j = \exp\left\{2\pi i\left(\frac{\Delta_j}{\Delta_n} + U\right)\right\}$, where U is a uniform random variable in $[0, 1)$.

It is a consequence of Proposition 3.11 that if $\{U_n\}_{n\in\mathbb{N}}$ is a sequence of Haar-distributed random unitary matrices, then $|\operatorname{Tr}(U_n)|^2$ converges to an exponential random variable with mean 1. In fact, this convergence happens very quickly: it follows from work of Johansson [60] that if $U \in \mathbb{U}(n)$ is distributed according to Haar measure, then for $t \geq 0$,

$$\left|\mathbb{P}[|\operatorname{Tr}(U_n)|^2 \geq t] - e^{-t}\right| \leq cn^{-\delta n}$$

for some universal constants c and δ. Coram and Diaconis computed the corresponding norm-squared "traces" of the wrapped zeta data; the comparison with the random matrix prediction is given in Figure 7.3.

A different global test of the wrapped zeta data has to do with the covariance structure of the counting function. Fix $\theta \in [0, 1]$, and let $I(\theta)$ be the quarter-

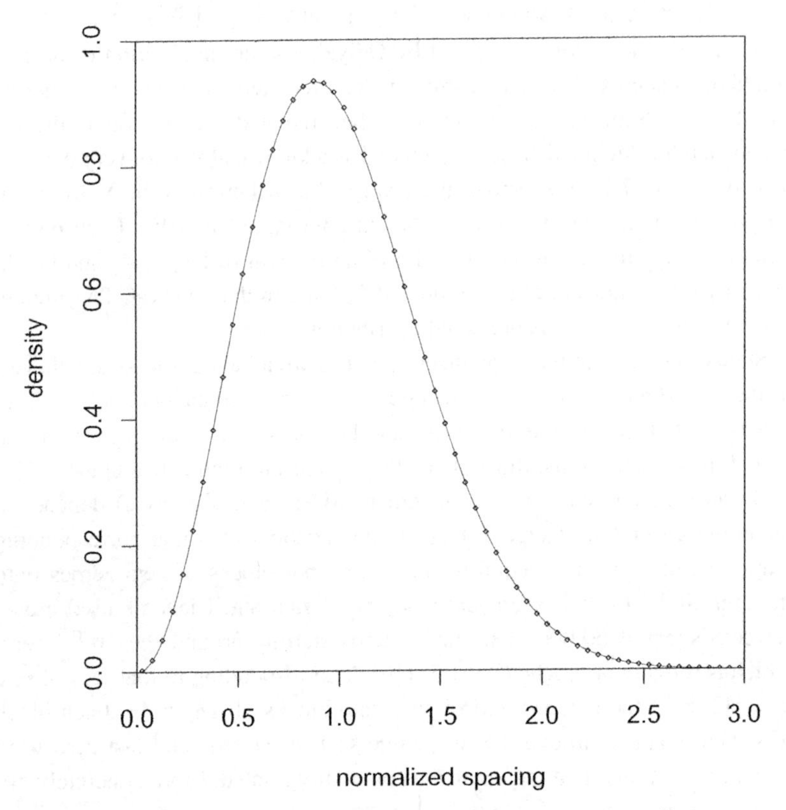

Figure 7.2 (Normalized) Neighbor spacings of zeta zeroes around the 10^{23} zero and the random matrix prediction. *Figure courtesy of Andrew Odlyzko.*

circle arc from $e^{2\pi i\theta}$ to $e^{2\pi i(\theta+\frac{1}{4})}$. Let U be a random matrix in $\mathbb{U}(n)$, and let $A(\theta)$ be the number of eigenvalues of U in $I(\theta)$. Finally, let

$$R(\theta) = \mathrm{corr}(A(\theta), A(0)).$$

An analytic expression for the density of $R(\theta)$ was found in [17]; the comparison with empirical correlations of the wrapped zeta deta is shown in Figure 7.4.

These pictures neatly convey the message of the extensive testing presented in [24], that numerical and statistical testing bear out the connection between zeta zeroes and random matrix eigenvalues amazingly well.

Finally, recall the conjecture of Keating and Snaith: that the distribution of values of the characteristic polynomial of an $n \times n$ random unitary matrix is a good model for the distribution of values of ζ at height T, where

$$n = \log\left(\frac{T}{2\pi}\right).$$

The analytic evidence presented for the conjecture in Section 7.2 is the agreement between Selberg's central limit theorem for the logarithm of the zeta function and Keating and Snaith's central limit theorem for the logarithm of the characteristic polynomial of a random unitary matrix. The numerical data here are striking: there random matrix prediction of the value distribution of $\log\zeta\left(\frac{1}{2}+it\right)$ by the distribution of $\log(Z(U,0))$ for a random matrix U is better than the Gaussian approximation given by the Selberg central limit theorem! In the figures below (reproduced from [65]), the value distribution for the real and imaginary parts of $\log(Z(U,0)$ are computed for U distributed according to Haar measure in $\mathbb{U}(42)$, and compared with value distributions (computed by Odlyzko) of the real and imaginary parts of $\log\zeta\left(\frac{1}{2}+it\right)$ for t near the $10^{20\text{th}}$ zero ($t \approx 1.5\times 10^{19}$). The Gaussian distributions predicted by the two central limit theorems are also plotted for comparison.

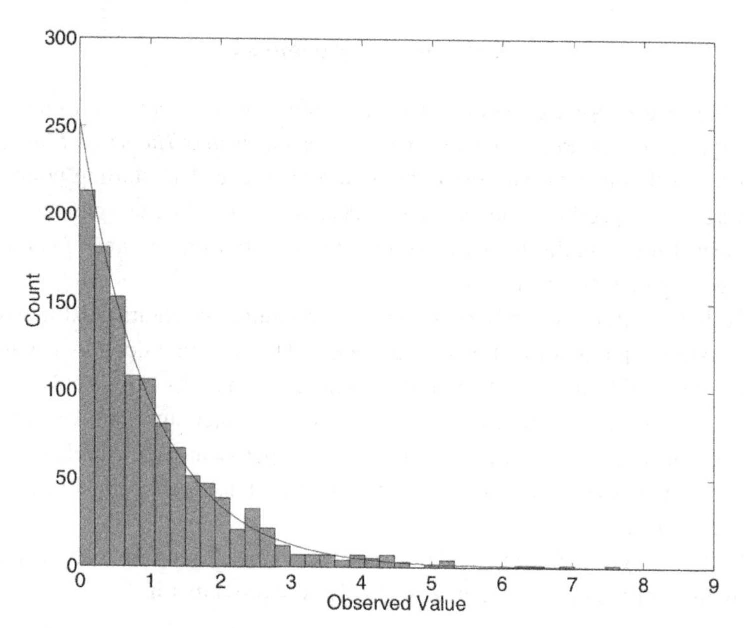

Figure 7.3 Empirical distribution of norm-squared "traces" of the wrapped zeta data and exponential density. *Reprinted by permission from IOP Publishing.*

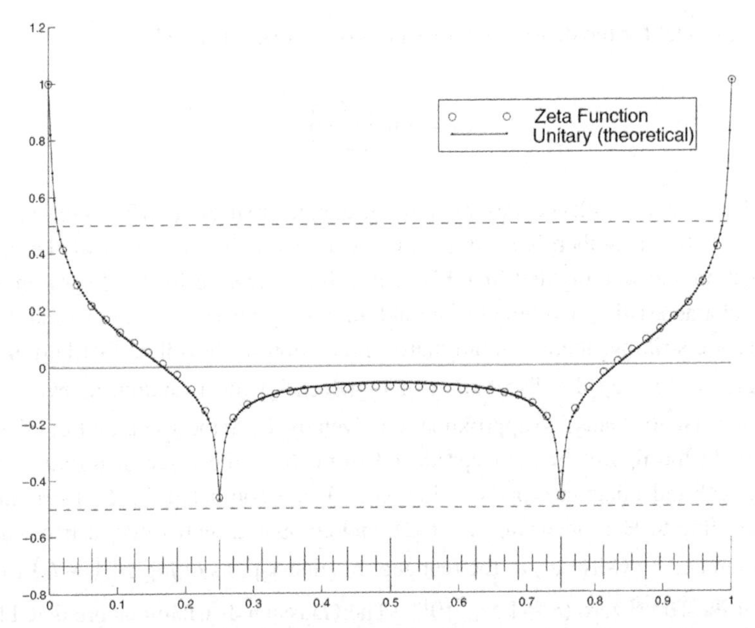

Figure 7.4 Estimated empirical values of $R(\theta)$ for the wrapped zeta data and the random matrix prediction. *Reprinted by permission from IOP Publishing.*

Notes and References

For the reader looking to delve into the random matrix approach to number theory, the volume *Recent Perspectives in Random Matrix Theory and Number Theory* [80] edited by Mezzadri and Snaith is an excellent starting point; its lectures were specifically intended to be accessible to researchers coming from number theory and those coming from random matrix theory (and to students just getting their feet wet in both!).

Following Montgomery's work on pair correlations, the natural next step was to consider k-point correlations for general k. This was first done by Rudnick and Sarnak [92] (in the context of more general L-functions, with the Riemann zeta function as a special case). This problem was taken up more recently by Conrey and Snaith [23], who introduced a new approach to the correlations on the random matrix side, which allowed for a more transparent comparison with the zeta zeroes.

In [56], Hughes, Keating, and O'Connell extended Theorem 7.4 by considering the entire random process $Z(U,\theta)$. They showed that if

$$Y_n(\theta) = \frac{1}{\sigma} \operatorname{Im} \log(Z(,\theta))$$

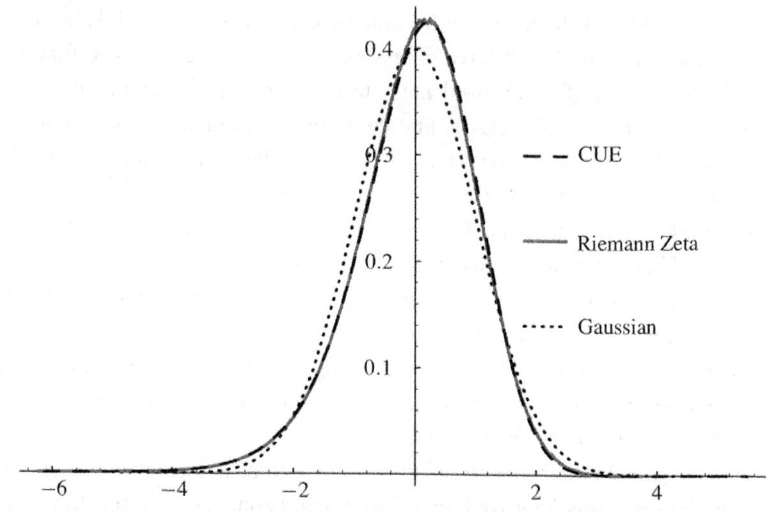

Figure 7.5 The value distributions of $\mathrm{Re}\log(Z(U,0))$ with $n = 42$, $\mathrm{Re}\log\zeta\left(\frac{1}{2}+it\right)$ near the $10^{20\text{th}}$ zero, and a Gaussian density, all scaled to have variance 1. *Reprinted by permission from Springer Nature.*

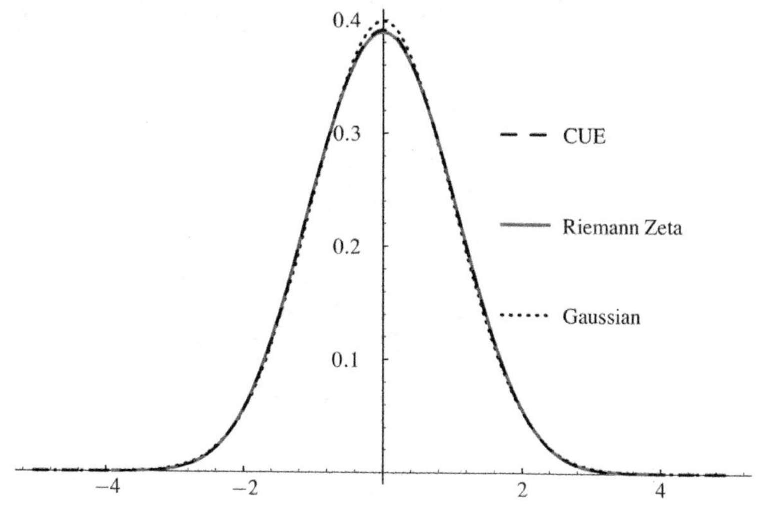

Figure 7.6 The value distributions of $\mathrm{Im}\log(Z(U,0))$ with $n = 42$, $\mathrm{Im}\log\zeta\left(\frac{1}{2}+it\right)$ near the $10^{20\text{th}}$ zero, and a Gaussian density, all scaled to have variance 1. *Reprinted by permission from Springer Nature.*

with $2\sigma^2 = \log(n)$, then the finite-dimensional distributions of $Y_n(\theta)$ converge to those of $Y(\theta)$, a centered Gaussian process with covariance function $\mathbb{E}Y(s)Y(t) = \mathbb{1}_{s=t}$. This allowed them to recover, via the argument principle, the covariances of the eigenvalue counting functions on arcs first found by Wieand [106]. In the same paper, they proved large deviations results for $\operatorname{Re}\log(Z(U,\theta))$ and $\operatorname{Im}\log(Z(U,\theta))$

In remarkable recent work, Chhaïbi, Najnudel, and Nikeghbali [19] have gone considerably further, showing that, after rescaling, the characteristic polynomial of a random unitary matrix converges almost surely to a random analytic function whose zeroes form a determinantal point process on the real line, governed by the sine kernel. This in turn suggested new conjectures on the value distribution of the zeta function, in an extension of the ideas that lead to the Keating–Snaith moment conjecture.

Finally, although the connection between zeta zeros and eigenvalues of random matrices remains unproved, there is a proven connection in the "function field" case; i.e., for zeta functions for curves over finite fields. Katz and Sarnak [62] showed that the zeros of almost all such zeta functions are described by the random matrix model. See also the expository article [63].

References

[1] Greg W. Anderson, Alice Guionnet, and Ofer Zeitouni. *An Introduction to Random Matrices*, volume 118 of *Cambridge Studies in Advanced Mathematics*. Cambridge University Press, Cambridge, 2010.

[2] Cécile Ané, Sébastien Blachère, Djalil Chafaï, Pierre Fougères, Ivan Gentil, Florent Malrieu, Cyril Roberto, and Grégory Scheffer. *Sur les Inégalités de Sobolev Logarithmiques*, volume 10 of *Panoramas et Synthèses [Panoramas and Syntheses]*. Société Mathématique de France, Paris, 2000. With a preface by Dominique Bakry and Michel Ledoux.

[3] Milla Anttila, Keith Ball, and Irini Perissinaki. The central limit problem for convex bodies. *Trans. Amer. Math. Soc.*, 355(12):4723–4735, 2003.

[4] Shiri Artstein-Avidan, Apostolos Giannopoulos, and Vitali D. Milman. *Asymptotic Geometric Analysis. Part I*, volume 202 of *Mathematical Surveys and Monographs*. American Mathematical Society, Providence, RI, 2015.

[5] Guillaume Aubrun, Stanisław Szarek, and Elisabeth Werner. Hastings's additivity counterexample via Dvoretzky's theorem. *Comm. Math. Phys.*, 305(1):85–97, 2011.

[6] Z. D. Bai. Methodologies in spectral analysis of large-dimensional random matrices, a review. *Statist. Sinica*, 9(3):611–677, 1999. With comments by G. J. Rodgers and Jack W. Silverstein; and a rejoinder by the author.

[7] Jinho Baik and Eric M. Rains. Algebraic aspects of increasing subsequences. *Duke Math. J.*, 109(1):1–65, 2001.

[8] D. Bakry and Michel Émery. Diffusions hypercontractives. In *Séminaire de probabilités, XIX, 1983/84*, volume 1123 of *Lecture Notes in Math.*, pages 177–206. Springer, Berlin, 1985.

[9] Dominique Bakry, Ivan Gentil, and Michel Ledoux. *Analysis and Geometry of Markov Diffusion Operators*, volume 348 of *Grundlehren der Mathematischen Wissenschaften [Fundamental Principles of Mathematical Sciences]*. Springer, Cham, 2014.

[10] Christian Berg, Jens Peter Reus Christensen, and Paul Ressel. *Harmonic Analysis On Semigroups: Theory of Positive Definite and Related Functions*, volume 100 of *Graduate Texts in Mathematics*. Springer-Verlag, New York, 1984.

[11] Rajendra Bhatia. *Matrix Analysis*, volume 169 of *Graduate Texts in Mathematics*. Springer-Verlag, New York, 1997.

[12] Gordon Blower. *Random Matrices: High Dimensional Phenomena*, volume 367 of *London Mathematical Society Lecture Note Series*. Cambridge University Press, Cambridge, 2009.

[13] Stéphane Boucheron, Gábor Lugosi, and Pascal Massart. *Concentration Inequalities: A Nonasymptotic Theory of Independence*. Oxford University Press, Oxford, 2013.

[14] Silouanos Brazitikos, Apostolos Giannopoulos, Petros Valettas, and Beatrice-Helen Vritsiou. *Geometry of Isotropic Convex Bodies*, volume 196 of *Mathematical Surveys and Monographs*. American Mathematical Society, Providence, RI, 2014.

[15] Ulrich Brehm and Jürgen Voigt. Asymptotics of cross sections for convex bodies. *Beiträge Algebra Geom.*, 41(2):437–454, 2000.

[16] Daniel Bump. *Lie Groups*, volume 225 of *Graduate Texts in Mathematics*. Springer, New York, second edition, 2013.

[17] Daniel Bump, Persi Diaconis, and Joseph B. Keller. Unitary correlations and the Fejér kernel. *Math. Phys. Anal. Geom.*, 5(2):101–123, 2002.

[18] Sourav Chatterjee and Elizabeth Meckes. Multivariate normal approximation using exchangeable pairs. *ALEA Lat. Am. J. Probab. Math. Stat.*, 4:257–283, 2008.

[19] Réda Chhaïbi, Joseph Najnudel, and Ashkan Nikeghbali. The circular unitary ensemble and the Riemann zeta function: the microscopic landscape and a new approach to ratios. *Invent. Math.*, 207(1):23–113, 2017.

[20] Benoît Collins. Moments and cumulants of polynomial random variables on unitary groups, the Itzykson-Zuber integral, and free probability. *Int. Math. Res. Not.*, (17):953–982, 2003.

[21] Benoît Collins and Piotr Śniady. Integration with respect to the Haar measure on unitary, orthogonal and symplectic group. *Comm. Math. Phys.*, 264(3):773–795, 2006.

[22] Benoît Collins and Michael Stolz. Borel theorems for random matrices from the classical compact symmetric spaces. *Ann. Probab.*, 36(3):876–895, 2008.

[23] J. B. Conrey and N. C. Snaith. In support of n-correlation. *Comm. Math. Phys.*, 330(2):639–653, 2014.

[24] Marc Coram and Persi Diaconis. New tests of the correspondence between unitary eigenvalues and the zeros of Riemann's zeta function. *J. Phys. A*, 36(12):2883–2906, 2003.

[25] S. Dallaporta. Eigenvalue variance bounds for Wigner and covariance random matrices. *Random Matrices Theory Appl.*, 1(3):1250007, 28, 2012.

[26] S. Dallaporta. Eigenvalue variance bounds for covariance matrices. *Markov Process. Related Fields*, 21(1):145–175, 2015.

[27] Anthony D'Aristotile, Persi Diaconis, and Charles M. Newman. Brownian motion and the classical groups. In *Probability, Statistics and their Applications: Papers in Honor of Rabi Bhattacharya*, volume 41 of *IMS Lecture Notes Monogr. Ser.*, pages 97–116. Inst. Math. Statist., Beachwood, OH, 2003.

[28] Kenneth R. Davidson and Stanislaw J. Szarek. Local operator theory, random matrices and Banach spaces. In *Handbook of the Geometry of Banach Spaces, Vol. I*, pages 317–366. North-Holland, Amsterdam, 2001.

[29] A. P. Dawid. Some matrix-variate distribution theory: notational considerations and a Bayesian application. *Biometrika*, 68(1):265–274, 1981.

[30] Amir Dembo and Ofer Zeitouni. *Large Deviations Techniques and Applications*, volume 38 of *Applications of Mathematics (New York)*. Springer-Verlag, New York, second edition, 1998.

[31] Persi Diaconis. *Group Representations in Probability and Statistics*, volume 11 of *Institute of Mathematical Statistics Lecture Notes—Monograph Series*. Institute of Mathematical Statistics, Hayward, CA, 1988.

[32] Persi Diaconis and Steven N. Evans. Linear functionals of eigenvalues of random matrices. *Trans. Amer. Math. Soc.*, 353(7):2615–2633, 2001.

[33] Persi Diaconis and Peter J. Forrester. Hurwitz and the origins of random matrix theory in mathematics. *Random Matrices Theory Appl.*, 6(1):1730001, 26, 2017.

[34] Persi Diaconis and David Freedman. Asymptotics of graphical projection pursuit. *Ann. Statist.*, 12(3):793–815, 1984.

[35] Persi Diaconis and David Freedman. A dozen de Finetti-style results in search of a theory. *Ann. Inst. H. Poincaré Probab. Statist.*, 23(2, suppl.):397–423, 1987.

[36] Persi Diaconis and Mehrdad Shahshahani. On the eigenvalues of random matrices. *J. Appl. Probab.*, 31A:49–62, 1994.

[37] Persi W. Diaconis, Morris L. Eaton, and Steffen L. Lauritzen. Finite de Finetti theorems in linear models and multivariate analysis. *Scand. J. Statist.*, 19(4):289–315, 1992.

[38] Christian Döbler and Michael Stolz. Stein's method and the multivariate CLT for traces of powers on the classical compact groups. *Electron. J. Probab.*, 16(86): 2375–2405, 2011.

[39] Richard M. Dudley. *Real Analysis and Probability*. The Wadsworth & Brooks/Cole Mathematics Series. Wadsworth & Brooks/Cole Advanced Books & Software, Pacific Grove, CA, 1989.

[40] Aryeh Dvoretzky. Some results on convex bodies and Banach spaces. In *Proc. Internat. Sympos. Linear Spaces (Jerusalem, 1960)*, pages 123–160. Jerusalem Academic Press, Jerusalem; Pergamon, Oxford, 1961.

[41] Morris L. Eaton. *Group Invariance Applications in Statistics*, volume 1 of *NSF-CBMS Regional Conference Series in Probability and Statistics*. Institute of Mathematical Statistics, Hayward, CA; American Statistical Association, Alexandria, VA, 1989.

[42] Alan Edelman and N. Raj Rao. Random matrix theory. *Acta Numer.*, 14:233–297, 2005.

[43] B. Fleury, O. Guédon, and G. Paouris. A stability result for mean width of L_p-centroid bodies. *Adv. Math.*, 214(2):865–877, 2007.

[44] P. J. Forrester. *Log-Gases and Random Matrices*, volume 34 of *London Mathematical Society Monographs Series*. Princeton University Press, Princeton, NJ, 2010.

[45] P. Frankl and H. Maehara. The Johnson-Lindenstrauss lemma and the sphericity of some graphs. *J. Combin. Theory Ser. B*, 44(3):355–362, 1988.

[46] Jason Fulman. Stein's method, heat kernel, and traces of powers of elements of compact Lie groups. *Electron. J. Probab.*, 17(66): 16, 2012.

[47] William Fulton and Joe Harris. *Representation Theory*, volume 129 of *Graduate Texts in Mathematics*. Springer-Verlag, New York, 1991.

[48] Yehoram Gordon. Some inequalities for Gaussian processes and applications. *Israel J. Math.*, 50(4):265–289, 1985.

[49] M. Gromov. Paul lévy isoperimetric inequality. I.H.E.S. preprint, 1980.

[50] M. Gromov and V. D. Milman. A topological application of the isoperimetric inequality. *Amer. J. Math.*, 105(4):843–854, 1983.

[51] Fumio Hiai and Dénes Petz. A large deviation theorem for the empirical eigenvalue distribution of random unitary matrices. *Ann. Inst. H. Poincaré Probab. Statist.*, 36(1):71–85, 2000.

[52] Fumio Hiai and Dénes Petz. *The Semicircle Law, Free Random Variables and Entropy*, volume 77 of *Mathematical Surveys and Monographs*. American Mathematical Society, Providence, RI, 2000.

[53] Roger A. Horn and Charles R. Johnson. *Topics in Matrix Analysis*. Cambridge University Press, Cambridge, 1994. Corrected reprint of the 1991 original.

[54] Roger A. Horn and Charles R. Johnson. *Matrix Analysis*. Cambridge University Press, Cambridge, second edition, 2013.

[55] J. Ben Hough, Manjunath Krishnapur, Yuval Peres, and Bálint Virág. Determinantal processes and independence. *Probab. Surv.*, 3:206–229, 2006.

[56] C. P. Hughes, J. P. Keating, and Neil O'Connell. On the characteristic polynomial of a random unitary matrix. *Comm. Math. Phys.*, 220(2):429–451, 2001.

[57] C. P. Hughes and Z. Rudnick. Mock-Gaussian behaviour for linear statistics of classical compact groups. *J. Phys. A*, 36(12):2919–2932, 2003. Random matrix theory.

[58] T. Jiang and Y. Ma. Distances between random orthogonal matrices and independent normals. *Preprint*, 2017. https://arxiv.org/abs/1704.05205.

[59] Tiefeng Jiang. How many entries of a typical orthogonal matrix can be approximated by independent normals? *Ann. Probab.*, 34(4):1497–1529, 2006.

[60] Kurt Johansson. On random matrices from the compact classical groups. *Ann. of Math. (2)*, 145(3):519–545, 1997.

[61] William B. Johnson and Joram Lindenstrauss. Extensions of Lipschitz mappings into a Hilbert space. In *Conference in modern analysis and probability (New Haven, Conn., 1982)*, volume 26 of *Contemp. Math.*, pages 189–206. American Mathematical Society, Providence, RI, 1984.

[62] Nicholas M. Katz and Peter Sarnak. *Random Matrices, Frobenius Eigenvalues, and Monodromy*, volume 45 of *American Mathematical Society Colloquium Publications*. American Mathematical Society, Providence, RI, 1999.

[63] Nicholas M. Katz and Peter Sarnak. Zeroes of zeta functions and symmetry. *Bull. Amer. Math. Soc. (N.S.)*, 36(1):1–26, 1999.

[64] J. P. Keating, F. Mezzadri, and B. Singphu. Rate of convergence of linear functions on the unitary group. *J. Phys. A*, 44(3):035204, 27, 2011.

[65] J. P. Keating and N. C. Snaith. Random matrix theory and $\zeta(1/2+it)$. *Comm. Math. Phys.*, 214(1):57–89, 2000.

[66] R. C. King. Modification rules and products of irreducible representations of the unitary, orthogonal, and symplectic groups. *J. Mathematical Phys.*, 12:1588–1598, 1971.

[67] B. Klartag. A central limit theorem for convex sets. *Invent. Math.*, 168(1):91–131, 2007.

[68] B. Klartag. Power-law estimates for the central limit theorem for convex sets. *J. Funct. Anal.*, 245(1):284–310, 2007.

[69] N. S. Landkof. *Foundations of Modern Potential Theory*. Springer-Verlag, New York-Heidelberg, 1972. Translated from the Russian by A. P. Doohovskoy, Die Grundlehren der mathematischen Wissenschaften, Band 180.

[70] Michel Ledoux. Concentration of measure and logarithmic Sobolev inequalities. In *Séminaire de Probabilités, XXXIII*, volume 1709 of *Lecture Notes in Math.*, pages 120–216. Springer, Berlin, 1999.

[71] Michel Ledoux. *The Concentration of Measure Phenomenon*, volume 89 of *Mathematical Surveys and Monographs*. American Mathematical Society, Providence, RI, 2001.

[72] Dudley E. Littlewood. *The Theory of Group Characters and Matrix Representations of Groups*. Oxford University Press, New York, 1940.

[73] Odile Macchi. The coincidence approach to stochastic point processes. *Advances in Appl. Probability*, 7:83–122, 1975.

[74] E. Meckes and M. Meckes. Rates of convergence for empirical spectral measures: A soft approach. In E. Carlen, M. Madiman, and E. Werner, editors, *Convexity and Concentration*, The IMA Volumes in Mathematics and its Applications, pages 157–181. Springer-Verlag, New York, 2017.

[75] Elizabeth Meckes. Linear functions on the classical matrix groups. *Trans. Amer. Math. Soc.*, 360(10):5355–5366, 2008.

[76] Elizabeth Meckes. Projections of probability distributions: a measure-theoretic Dvoretzky theorem. In *Geometric aspects of functional analysis*, volume 2050 of *Lecture Notes in Math.*, pages 317–326. Springer, Heidelberg, 2012.

[77] Elizabeth S. Meckes and Mark W. Meckes. Concentration and convergence rates for spectral measures of random matrices. *Probab. Theory Related Fields*, 156(1-2):145–164, 2013.

[78] Elizabeth S. Meckes and Mark W. Meckes. Spectral measures of powers of random matrices. *Electron. Commun. Probab.*, 18(78): 13, 2013.

[79] Madan Lal Mehta. *Random Matrices*, volume 142 of *Pure and Applied Mathematics (Amsterdam)*. Elsevier/Academic Press, Amsterdam, third edition, 2004.

[80] F. Mezzadri and N. C. Snaith, editors. *Recent Perspectives in Random Matrix Theory and Number Theory*, volume 322 of *London Mathematical Society Lecture Note Series*. Cambridge University Press, Cambridge, 2005.

[81] Francesco Mezzadri. How to generate random matrices from the classical compact groups. *Notices Amer. Math. Soc.*, 54(5):592–604, 2007.

[82] V. D. Milman. A new proof of A. Dvoretzky's theorem on cross-sections of convex bodies. *Funkcional. Anal. i Priložen.*, 5(4):28–37, 1971.

[83] Vitali D. Milman and Gideon Schechtman. *Asymptotic Theory of Finite-Dimensional Normed Spaces*, volume 1200 of *Lecture Notes in Mathematics*. Springer-Verlag, Berlin, 1986. With an appendix by M. Gromov.

[84] H. L. Montgomery. The pair correlation of zeros of the zeta function. pages 181–193, 1973.

[85] Robb J. Muirhead. *Aspects of Multivariate Statistical Theory*. John Wiley & Sons, Inc., New York, 1982. Wiley Series in Probability and Mathematical Statistics.

[86] Angela Pasquale. Weyl's integration formula for U(N). Based on an introductory lecture delivered at the DMV Seminar "The Riemann Zeta Function and Random Matrix Theory," October, 2000, Oberwolfach, Germany. Available online at http://www.math.tau.ac.il/~rudnick/dmv/Weyl.ps.

[87] Leonid Pastur and Mariya Shcherbina. *Eigenvalue Distribution of Large Random Matrices*, volume 171 of *Mathematical Surveys and Monographs*. American Mathematical Society, Providence, RI, 2011.

[88] E. M. Rains. High powers of random elements of compact Lie groups. *Probab. Theory Related Fields*, 107(2):219–241, 1997.

[89] E. M. Rains. Increasing subsequences and the classical groups. *Electron. J. Combin.*, 5:Research Paper 12, 9, 1998.

[90] Eric M. Rains. Images of eigenvalue distributions under power maps. *Probab. Theory Related Fields*, 125(4):522–538, 2003.

[91] Arun Ram. Characters of Brauer's centralizer algebras. *Pacific J. Math.*, 169(1):173–200, 1995.

[92] Zeév Rudnick and Peter Sarnak. Zeros of principal L-functions and random matrix theory. *Duke Math. J.*, 81(2):269–322, 1996.

[93] Gideon Schechtman. A remark concerning the dependence on ϵ in Dvoretzky's theorem. In *Geometric aspects of functional analysis (1987–88)*, volume 1376 of *Lecture Notes in Math.*, pages 274–277. Springer, Berlin, 1989.

[94] A. Soshnikov. Determinantal random point fields. *Uspekhi Mat. Nauk*, 55(5(335)):107–160, 2000.

[95] Alexander Soshnikov. The central limit theorem for local linear statistics in classical compact groups and related combinatorial identities. *Ann. Probab.*, 28(3):1353–1370, 2000.

[96] Charles. Stein. The accuracy of the normal approximation to the distribution of the traces of powers of random orthogonal matrices. Technical Report No. 470, Stanford University Department of Statistics, 1995.

[97] Kathryn Stewart. Total variation approximation of random orthogonal matrices by gaussian matrices. *Journal of Theoretical Probability*, to appear. https://arxiv.org/abs/1704.06641.

[98] Michael Stolz. On the Diaconis-Shahshahani method in random matrix theory. *J. Algebraic Combin.*, 22(4):471–491, 2005.

[99] V. N. Sudakov. Typical distributions of linear functionals in finite-dimensional spaces of high dimension. *Dokl. Akad. Nauk SSSR*, 243(6):1402–1405, 1978.

[100] Michel Talagrand. *The Generic Chaining*. Springer Monographs in Mathematics. Springer-Verlag, Berlin, 2005. Upper and lower bounds of stochastic processes.

[101] Santosh S. Vempala. *The Random Projection Method*, volume 65 of *DIMACS Series in Discrete Mathematics and Theoretical Computer Science*. American Mathematical Society, Providence, RI, 2004. With a foreword by Christos H. Papadimitriou.

[102] Cédric Villani. *Optimal Transport: Old and New*, volume 338 of *Grundlehren der Mathematischen Wissenschaften [Fundamental Principles of Mathematical Sciences]*. Springer-Verlag, Berlin, 2009.

[103] Frank W. Warner. *Foundations of Differentiable Manifolds and Lie Groups*, volume 94 of *Graduate Texts in Mathematics*. Springer-Verlag, New York-Berlin, 1983. Corrected reprint of the 1971 edition.

[104] Fred B. Weissler. Logarithmic Sobolev inequalities and hypercontractive estimates on the circle. *J. Funct. Anal.*, 37(2):218–234, 1980.

[105] Hermann Weyl. *The Classical Groups. Their Invariants and Representations.* Princeton University Press, Princeton, NJ, 1939.

[106] K. Wieand. Eigenvalue distributions of random unitary matrices. *Probab. Theory Related Fields*, 123(2):202–224, 2002.

[107] P. Wojtaszczyk. *Banach Spaces for Analysts*, volume 25 of *Cambridge Studies in Advanced Mathematics*. Cambridge University Press, Cambridge, 1991.

Index